Lecture Notes in Mathematics

Edited by A. Dold and B. Eckmann

664

Algebraic and Geometric Topology

Proceedings of a Symposium held at Santa Barbara in honor of Raymond L. Wilder, July 25–29, 1977

Edited by Kenneth C. Millett

Springer-Verlag
Berlin Heidelberg New York 1978

Editor
Kenneth C. Millett
Department of Mathematics
University of California
Santa Barbara, CA 93106/USA

Library of Congress Cataloging in Publication Data

Algebraic and Geometric Topology Symposium, University of California, Santa Barbara, 1977.
Algebraic and geometric topology.

(Lecture notes in mathematics ; 664)
Sponsored by the National Science Foundation and the University of California, Santa Barbara.
Bibliography: p.
Includes index.
1. Algebraic topology--Congresses. 2. Manifolds (Mathematics)--Congresses. 3. Wilder, Raymond Louis, 1896- I. Wilder, Raymond Louis, 1896- II. Millett, Kenneth C. III. United States. National Science Foundation. IV. California. University,
QA3.L28 no. 664 [QA612] 510'.8s [514'.2] 78-15091

AMS Subject Classifications (1970): 55-00, 55-02, 57-00 57-02, 57-03

ISBN 3-540-08920-9 Springer-Verlag Berlin Heidelberg New York
ISBN 0-387-08920-9 Springer-Verlag New York Heidelberg Berlin

Printing and binding: Beltz Offsetdruck, Hemsbach/Bergstr.
2141/3140-543210

Dedicated to

RAYMOND LOUIS WILDER

1896 -

It is a rare privilege to be honored, at my age, by younger colleagues whose interests, while paralleling those which dominated my thoughts during the second quarter of this century, have extended so much further into the frontiers of Topology.

In an address before the American Association for the Advancement of Science 25 years ago*, I remarked: "I envy those young men who are only on the threshold of their mathematical careers, for they will be posssessed of powers that will put their elders to shame. I have already noticed ... evidence of greater powers among recent recruits to the ranks of mathematical research. The future will no doubt see an increase in this... ." The papers in this volume contribute further justification for my earlier observations. To their authors I offer profound thanks, both personal for myself, and impersonal for mathematics, for the results which they have been obtaining.

I am also deeply indebted to Professor Kenneth Millett and his co-sponsors, the University of California, Santa Barbara and the National Science Foundation, for the preparation of this volume and the conference of July, 1977, at which papers published herein were presented.

R. L. Wilder
Santa Barbara, California
January 27, 1978

---

*The Origin and Growth of Mathematical Concepts, Bull. Amer. Math. Soc. 59 (1953), 423-448.

PREFACE

This volume contains the proceedings of the Algebraic and Geometric Topology Symposium held in honor of Professor Raymond L. Wilder at the University of California, Santa Barbara, under the joint sponsorship of the National Science Foundation and the University of California.

The Symposium was organized to have a special emphasis on the theory of generalized manifolds. This choice was motivated by the desire to honor the scholarly achievements of Professor Wilder on the occasion of his 80th birthday by emphasizing an important area of current research in topology with which his name is associated. Thus the theory of generalized manifolds, with its connections to questions of decomposition spaces, the double suspension problem, CE-maps, questions of structures on manifolds, and the theory of transformation groups made it an ideal focus of the Symposium.

These topics were well represented among the lectures presented during the Symposium and papers submitted by authors desiring to join in a volume honoring Wilder. There are also three papers given at the Symposium which have a uniquely historical perspective, indicating Wilder's influence in various areas of mathematics as well as pointing to some problems of continuing interest.

I wish to thank the National Science Foundation and the University of California for their sponsorship of the Symposium, my colleagues at the University of California and their families for their encouragement and assistance, and to all the participants who joined in creating a mathematical and colleagual event appropriate to the honor of Ray Wilder.

This volume was typed by Michelle Dunn, of the Institute for Interdisciplinary Applications of Algebra and Combinatorics, and Marianne Braun, Ruth Hillard and Jill Weaver of the Department of Mathematics at the University of California, Santa Barbara and I wish to thank them for their patient cooperation during the production of this manuscript.

Finally I wish to thank all the participants in the Symposium for their assistance and expression of concern during the course of the Santa Barbara Fire of 1977 which was unfortunately scheduled for the same week as the Symposium.

Kenneth C. Millett
Santa Barbara, California
January 31, 1978

SCHEDULE OF LECTURES

MONDAY, JULY 25	Edwards: CE Maps of Manifolds
	Cohen: Stieltjes Derivative
	Hirsch: Flat Manifolds and Cohomology of Groups
TUESDAY, JULY 26	Hatcher: Smale Conjecture
	Smale: Hilbert's 16th Problem
	Raymond: Diffeotopy classes of diffeomorphisms of certain 3-dimensional homology spheres
	Michelsohn: Clifford Bundles on Manifolds
	Jones: Wilder on Connectedness
	Parks: Self-Homotopy Equivalences
WEDNESDAY, JULY 27	Ferry: ϵ-maps
	Stern: Homology 3-spheres and triangulating manifolds
	Morgan: Poincare duality
	Moise: Statically tame periodic homeomorphisms of compact connected 3-manifolds
	Brown: A characterization of inner automorphisms
THURSDAY, JULY 28	Raymond: Discussion of Wilder's Topological Interests: solved and unsolved problems, new developments,...
	Farrell: The topological-Euclidean space form problem
	Brumfiel: Representing homology classes by PL manifolds
	McCrory: Intersection of cycles in stratified spaces
	Curtis: HP^{∞}
FRIDAY, JULY 29	Orlik: Finite complex reflection groups
	Larmore: Single obstructions to embedding
	Scharlemann: Approximating smooth CE maps

PARTICIPANTS

Glen E. Bredon, Rutgers University

Robert Brown, University of California, Los Angeles

Gregory Brumfiel, Stanford University

Gunnar Carlsson, Stanford University

Leon W. Cohen, University of Maryland

Marc Culler, University of California, Berkeley

Carlos Curley, Boston College

Morton Curtis, Rice University

Karl Heinz Dovermann, Rutgers University

Robert Edwards, University of California, Los Angeles

John Ernest, University of California, Santa Barbara

Tom Farrell, Institute for Advanced Study

Steve Ferry, University of Kentucky

Michael Freedman, University of California, San Diego

Herb Gindler, San Diego State University

Sue Goodman, University of North Carolina

Leroy Grant, University of California, Irvine

Jenny Harrison, University of California, Berkeley

Allen Hatcher, University of California, Los Angeles

Morris Hirsch, University of California, Berkeley

John Hocking, Michigan State University

Jozo Hunjic, University of California, Santa Barbara

F. Burton Jones, University of California, Riverside

Lowell Jones, SUNY, Stony Brook

Ralph M. Krause, National Science Foundation

Kyung Whan Kwun, Michigan State University

L. Larmore, California State College, Domingues Hills

T. Y. Lin, National Taiwan University

Clint McCrory, Brown University

Marie-Louise Michelsohn, University of California, Berkeley

Richard T. Miller, Michigan State University

Kenneth C. Millett, University of California, Santa Barbara

Edwin Moise, City University of New York, Queens College

John D. Moore, University of California, Santa Barbara

John Morgan, Columbia University

Peter Orlik, University of Wisconsin

James Parks, Howard University

D. H. Potts, California State University, Northridge

Frank Raymond, University of Michigan

Elmer Rees, St. Catherine's College, Oxford

H. Samelson, Stanford University

M. Scharlemann, University of California, Santa Barbara

Charles Seebeck, Michigan State University

Stephen Smale, University of California, Berkeley

Ronald Stern, University of Utah

P. Emery Thomas, University of California, Berkeley

Bruce Trace, University of California, Los Angeles

Karen Vogtmann, University of California, Berkeley

Lucille Whyburn, University of Texas

Raymond L. Wilder, University of California, Santa Barbara

R. Wong, University of California, Santa Barbara

TABLE OF CONTENTS

WILDER ON CONNECTEDNESS

by F. Burton Jones
University of California, Riverside

I shall begin in a sense at the beginning, i.e., when Fundamenta Mathematica first started. In 1921 Sierpinski showed [16] that there exists (in the plane) a point p and a point set N with no nondegenerate quasi-component such that $N \cup \{p\}$ although totally disconnected does contain a nondegenerate quasi-component. Recall that a quasi-component of a point set M is not necessarily connected but rather in no way is M the union of two mutually separated subsets each intersecting it. (In present day terminology one might say that a quasi-component of M is a subset of M which is maximal with respect to the property that no two of its points are separated in M by the empty set.) The same year Knaster and Kuratowski showed [7] that there exists a point p and a set N with no nondegenerate component (i.e., totally disconnected) such that $N \cup \{p\}$ is connected. But in this case N has nondegenerate quasi-components. (This is the famous "explosion point" example constructed in the Cantor fan by taking those points on accessible intervals with rational ordinates and irrational ordinates otherwise. The points on any interval form a quasi-component.) This raised the question: Is there such a set N with no nondegenerate quasi-components? In 1927 Wilder settled this question by constructing such an example - a rather complicated one [19].

I produced another in 1942 by constructing a discontinuous solution to the functional equation $f(x) + f(y) = f(x+y)$ with a connected plane graph [3]. Such graphs are dense in the plane. Obviously such a connected graph minus its intersection with a non-vertical, straight line is a set N with only degenerate quasi-components. But N plus "the point at infinity" is connected. And in 1956 Roberts mapped the rational points in Hilbert space homeomorphically into the (plane) Cantor fan so that no interval of the fan contained more than one image point [15]. Again quasi-components are degenerate. Nevertheless, this image plus the vertex of the fan is connected.

These last two examples, although still quite non-intuitive can in some sense be said "to exist in nature".

These weird connected sets (those with "explosion points") are punctiform, i.e., contain no nondegenerate continuum. In 1921 Moore gave an example of a connected, locally connected, nondegenerate connected set which contained no arc [14]. Wilder produced one which contained no nondegenerate continuum [20]. The connected discontinuous solution of $f(x) + f(y) = f(x+y)$ gives another [4]. Take the union of such a connected graph and all its rational vertical translates. This set is connected, locally connected and contains no nondegenerate continuum.

Today we may well wonder why all this interest in these strange sets. We now think of an arc as a compact connected space (usually Hausdorff or metric) which is separated by every point except possibly two. But in those days they were groping for the most general satisfactory characterization. Wilder, in particular, wanted to avoid using "closed" or "bounded" [preferably both] if at all possible, not just in this connection but in general.

Suppose a connected subset of a space (even the plane) is irreducibly connected between two points. Would it have to be an arc? [Far from it; the closed interval of a connected discontinuous f from $(0,0)$ to $(1,f(x))$ doesn't even contain an arc.] However, Whyburn showed in 1927 that it would be an arc <u>if</u> locally connected [18]. This was for the plane and a year later Wilder got the same result for subsets of E^m [21]. In fact, Wilder went further by showing that any connected, locally connected subset of E^m which is irredubily connected about a compact set is necessarily a compact continuum. [We define here, as then, that a continuum is a closed connected set, perhaps not compact. This applies throughout this paper.] About three years later Wilder discarded local connectivity and got the following: If M is a continuum, $a, b \in M$, and every $p \in M - \{a,b\}$ separates a from b, then M is an arc [22].

Wilder was interested in doing the same kind of thing for simple closed curves and as we shall see later, for all simple continuous curves - and, in fact, for hereditarily, locally connected, continua.

In 1924 Kline showed that a plane continuum which is not separated by any connected subset must be a simple closed curve [5] and that same year Kuratowski got the same conclusion in the compact case if no subcontinuum cuts [its complement not strongly connected] [10].

A little later Wilder defined a quasi-closed curve M as the union of two sets M_1 and M_2 irreducibly connected from a to b such that $M_1 - \{a,b\}$ and $M_2 - \{a,b\}$ are mutually separated. Every connected set M which is separated by no point but by every pair of points is a quasi-closed curve (and conversely) and if closed, M is a simple closed curve. Furthermore, if M is connected and locally connected, M is a simple closed curve if and only if M is a quasi-closed curve [23]. This led Wilder to conjecture that one might discard the above mutually separated requirement in the presence of local connectivity and still prove that M is a simple closed curve. In 1948 Bing, by using a scheme like the one I had used on $f(x) + f(y) = f(x+y)$, got two non-intersecting graphs (one with its y-axis tilted slightly from the vertical) whose union intersects every subcontinuum of the plane and is therefore locally connected. Adding $-\infty$ and $+\infty$ gives the counter-example [1].

In 1929 Wilder showed for locally compact continua, hereditary local connectivity is equivalent to every component of a G_δ-subset being either strongly connected or arcwise connected [23]. Forty years later, Mohler shows that for a compact continuum to be hereditarily locally connected one need only assume that every connected G_δ-subset is strongly connected [12].

But the most revealing and imaginative result along this line occurs in Wilder's thesis: A compact continuum is locally connected if and only if every component of an open set is strongly connected [24]. It was almost fifty years later (1970) that Mohler showed the same to be true if and only if every connected, open set is strongly connected (local compactness is sufficient) [13]. This result came long after Knaster had specifically raised the question in 1937. It does not generalize to the non-metric case.

Let us return to another question that arose along with "explosion points".

Kline showed (it's quite simple) that no connected set contains more than one "explosion point" (= dispersion point) [6]. Wilder defined a dispersion set as a subset D of a connected set whose complement is totally disconnected and, if no proper subset of D has this property, a primitive dispersion set. He proved among other things, that no connected set has more than one finite primitive dispersion set [25]. Not all connected sets have primitive dispersion sets [e.g., the real unit interval], although clearly every nondegenerate connected set has a dispersion set. But can a connected set have two (infinite) primitive dispersion sets? Wilder had such an example but it was in 3-space (*loc. cit.*). Knaster, intrigued with the notion, produced one in the plane [8]. As I recall, one was countable and the other uncountable. But Wilder raised the general question about which spaces *did* have dispersion sets - does the plane, for example? Mary Ellen Rudin (nee Estill) showed this to be the case by constructing a totally disconnected subset of the plane which was no longer so if one added to it any point of its complement [2].

Now it is true that if a connected set has a primitive dispersion set with more than two points in it, then the connected set is itself the union of two disjoint connected subsets. Why is this of interest? Knaster and Kuratowski defined a connected set which is NOT the union of two disjoint connected subsets to be *biconnected* [9]. A connected set with a degenerate dispersion set is an example of a biconnected set. So one cannot get a biconnected set without a dispersion point by going to dispersion sets with more than one point. Actually in 1922 Kuratowski raised the question in *Fundamenta*: Does every biconnected set have a dispersion point? [26].

At this point I must talk about two of Ray's students: P.M. Swingle and E.W. Miller. About 1931 Swingle introduced the notion of a widely connected set, i.e., a connected set in which every nondegenerate connected subset is dense in it [17]. Obviously such a set cannot contain a separating point, much less a connected subset with a dispersion point. But Swingle was not able to prove that some widely connected set was biconnected. So while this attempt at solving Kuratowski's problem failed, it was on the right track.

One weekend, while Ray was puttering around in the garden, Ed Miller drove up. After a brief greeting, Ed said (approximately):

"Didn't you offer a bottle of liquor for an example of a biconnected set with no dispersion point?"

Wilder thought, "I'd better be careful about this." Well, maybe he had made some such remark.

"What if you used the continuum hypothesis?"

"Oh, that's different. It might be worth a drink."

"Well, break out the bottle."

Published in 1937, [11] Ed had indeed constructed such a set in one of Swingle's widely connected sets! Even today no one has succeeded without the continuum hypothesis.

BIBLIOGRAPHY

[1] Bing, R.H., Solution of a problem of R.L. Wilder, Amer. Jour. Math. 70 (1948), 95-98.

[2] Estill, M.E., A primitive dispersion set of the plane, Duke Math. Jour. 19 (1952), 323-328.

[3] Jones, F.B., Connected and disconnected plane sets and the functional equation $f(x) + f(y) = f(x+y)$, Bull. Amer. Math. Soc. 48 (1942), 115-120.

[4] Jones, F.B., loc. cit.

[5] Kline, J.R., Closed connected sets which remain connected upon the removal of certain connected subsets, Fund. Math. 5 (1924), 3-10.

[6] Kline, J.R., A theorem concerning connected point sets (sic), Fund. Math. 3 (1922), 238-239.

[7] Knaster, B.;Kuratowski, C., Sur les ensembles connexes, Fund. Math. 2 (1921), 206-255.

[8] Knaster, B., Sur un probleme de M.R.L. Wilder, Fund. Math. 7 (1925), 191-197.

[9] Knaster, B.;Kuratowski, C., Sur les ensembles connexes, Fund. Math. 2 (1921), 206-255.

[10] Kuratowski, C., Contribution à l'étude de continus de Jordan, Fund. Math. 5 (1924), 112-122.

[11] Miller, E.W., Concerning biconnected sets, Fund. Math. 29 (1937), 123-133.

[12] Mohler, L., A note on hereditarily locally connected continua, Bull. de l'Acad. Polonaise des Sciences 17 (1969), 699-701.

[13] Mohler, L., A characterization of local connectedness for generalized continua, Coll. Math. 21 (1970), 81-85.

[14] Moore, R.L., A connected and regular point set which contains no arc, Bull. Amer. Math. Soc. 32 (1926), 331-332.

[15] Roberts, J.H., The rational points in Hilbert space, Duke Math. Jour. 23 (1956), 489-491.

[16] Sierpinski, W., Sur les ensembles connexes et non connexes, Fund. Math. 2 (1921), 81-95.

[17] Swingle, P.M., Two types of connected sets, Bull. Amer. Math. Soc. 37 (1931), 254-258.

[18] Whyburn, G.T., Concerning connected and regular point sets, Bull. Amer. Math. Soc. 33 (1927), 685-689.

[19] Wilder, R.L., A point set which has no true quasi-components, and which becomes connected upon the addition of a single point, Bull. Amer. Math. Soc. 33 (1927), 423-427.

[20] Wilder, R.L., A connected and regular point set which has no subcontinuum, Trans. Amer. Math. Soc. 29 (1927), 332-340.

[21] Wilder, R.L., On connected and regular point sets, Bull. Amer. Math. Soc. 34 (1928), 649-655.

[22] Wilder, R.L., Concerning simple continuous curves and related point sets, Amer. Jour. Math. 53 (1931), 39-55.

[23] Wilder, R.L., Characterizations of continuous curves that are perfectly continuous, Proc. Nat. Acad. Sci. 15 (1929), 614-621.

[24] Wilder, R.L., Concerning continuous curves, Fund. Math. 7 (1925), 340-377.

[25] Wilder, R.L., On the dispersion sets of connected point-sets (sic), Fund. Math. 6 (1924), 214-228.

[26] Problème de M. Kuratowski, Fund. Math. 3 (1922), p. 322, Prob. 19.

R. L. WILDER'S WORK ON GENERALIZED MANIFOLDS - AN APPRECIATION

Frank Raymond
University of Michigan

To R. L. Wilder

On November 3, 1896, R. L. Wilder discovered America. He grew up in Palmer, Massachusetts, and as a youth attended the schools of that town. His family was musical, and he played the cornet in the family orchestra at dances and fairs, as well as the piano. This taste for music making has never left him although he now usually sticks with the classics.

He entered Brown University in 1914, served in World War I, completed his bachelor's degree in 1920 and his master's degree in 1921. In the meantime he had made the most important decision of his life when he married the charming Una Greene. Together they raised a lovely family of 4 children and at last count are grandparents to 22 and great-grandparents to 6.

Those who know Ray and Una intimately are aware that above all these warm and gracious people live a beautiful and rich life which is alive with concern for the welfare of others. This humanism has suffused their entire scholarly lives and this aspect is deeply reflected in Professor Wilder's contributions to mathematics and to the mathematical community. Those who have had him as a teacher glow when they speak of their experiences. It makes little difference whether or not they were "star" pupils. All maintain that their contact with him was very rewarding and enriching and not limited to just mathematics. At the University of Michigan Wilder's wisdom is legendary. However, this legend did not always extend to the administration where one often found him at loggerheads with those who either through lack of courage or insight would trample on the purposes of a University.

In 1921 Ray and Una left New England for the University of Texas with the intention of continuing graduate work in actuarial mathematics. Ray liked "pure" mathematics and thought he would "squeeze some in" while at Texas. He asked

R. L. Moore if he could attend his course in analysis-situs. Of course Moore said "No, there is no way a person interested in actuarial mathematics could be able to do, let alone be really interested in, topology." Wilder was not deterred and persisted in his request. This determination was the "open sesame" to Moore's course. In two years Wilder received his Ph.D. under Moore.

He stayed an additional year at Austin as an instructor and then held an assistant professorship at Ohio State University from 1924 to 1926. When offered a position at the University of Michigan in 1926, he moved to Ann Arbor. He was appointed Research Professor at the University of Michigan in 1947, and in 1967 he became Emeritus Professor. Unfortunately, for us at Michigan, he then moved to a more congenial climate and has been an active participant in the intellectual life of the University of California at Santa Barbara ever since.

I would like to mention some of the numerous honors won by Professor Wilder over the years. He was a member of the Institute for Advanced Study, 1933-1934, held a Guggenheim Fellowship, 1940-1941, and held a number of visiting appointments at various universities. At Michigan he was appointed the Russel Lecturer in 1959, the University's highest honor. In 1939 he delivered the semi-centennial address of the American Math. Society entitled "The Sphere in Topology." In 1942 he delivered the A.M.S. Colloquium Lectures which resulted in publication of "Topology of Manifolds" in 1949. At the International Congress of Mathematicians, Cambridge, 1950, he delivered the address: "The Cultural Basis of Mathematics." In 1969 he delivered the Gibbs lecture of the A.M.S., "Trends and Social Implications of Research." He served in various capacities in the American Mathematical Society and was Vice President in 1950 and 1951 and President in 1955 and 1956. He was also President of the Mathematical Association of America in 1965 and 1966. I have appended a list of his published works up to 1977. This list is ever increasing. One can sense simply by a perusal of titles his wide scholarly and human interests. In this appreciation of Professor Wilder's published works I will concentrate only upon his contribution to the theory of generalized manifolds. Regretfully, I am leaving out his important contributions to other branches of topology, foundations of mathematics and the cultural evolution of mathematics.

A Unified Analysis-Situs

R. L. Wilder's first works in set-theoretic topology, well documented by Burton Jones' article of this symposium [J], were already well underway when the paper [A-1] of J. W. Alexander was studied in seminar at the University of Michigan. This very important paper, in which Alexander proved his now famous duality theorem, was instrumental in turning Wilder's interest towards manifold theory and the use of algebraic techniques. For us, equipped with the systematic machinery of algebraic topology, Alexander's theorem does not seem so overwhelming today. In fact, many graduate students can produce proofs of considerably more sophisticated versions of Alexander's theorem with ease and finesse. But we must remember at that time cohomology, relative homology, products, exact sequences, homotopy theory, etc. were still years in the future. Alexander had also produced his famous horned sphere [A-2] which meant the end of any hope for routine generalizations of plane topology to n-space. Yet the closed complementary domains of Alexander's sphere, while not both 3-cells, could not be distinguished from 3-cells by any homological means. It was this insight that led to Wilder's [W-1] and later to [W-2]. Here a converse of the Jordan-Separation theorem and characterization of generalized k-manifolds embedded in the sphere are obtained from homological conditions on the complementary domain(s).

In 1933 the Institute for Advanced Study was founded in Princeton. Roaming the corridors of old Fine Hall were many topologists including Veblen, Alexander, Lefschetz, Van Kampen, Tucker, Zippin and Wilder. Alexander, Čech, Vietoris, Alexandroff and Lefschetz had invented, or were inventing, various homology theories that could handle general spaces and their subsets. The notion of a generalized manifold was not unknown in the polyhedral category (Van Kampen, 1929, [V.K.]) and the formulation for topological spaces in more abstract homological terms was done in the early 1930's. The most significant papers seem to have been by V. Kampen [V.K.], Čech [C-1 through 5], Lefschetz [Le], Wilder [W-2], Alexandroff-Pontrjagin [A-P], Smith [Sm]. The first proofs of dualities for generalized manifolds appear in Čech [C-2] and Lefschetz [Le].

It is instructive to turn to Wilder's Symposium Lecture [W-3] delivered to The AMS in Chicago, 1932. Wilder had been concerned over the separation that had developed between the two schools of American topologists typified by the Texas (set theoretic or local) and Princeton (combinatoric or global) schools. Wilder was a successful "rebel" from the Texas school. It annoyed him to see criticism raised against unified methods. By combining the methods of both schools he had been able to obtain generalizations of theorems of the plane whose extensions to n-space by means of set-theoretic methods alone had heretofore been unsuccessful. He was not alone of course in his successes for Alexandroff, Čech and Lefschetz, to name a few, had no qualms about combining methods. But Wilder was brave to expose so much of his point of view to his contemporaries and to future generations, as he did in this Symposium Lecture. Wilder is very clear as to what his hopes, fears and aspirations for topology are. He does an excellent and fascinating job of summarizing and interpreting the then very active set-theoretic investigations in the plane in terms of the corresponding or the newly developed ideas of algebraic topology. To proceed to analogous theorems in higher dimensions he shows how one uses the ideas of higher connectedness, that is homology theory. Actually, he was criticizing the dogmatism of both schools. First, to ignore or deprecate certain problems because one's own, perhaps quite successful. methods do not seem to apply often leads to bitterness and many lost opportunities. Secondly, failure to accept certain unifying concepts can lead to sterility and isolation from the mainstreams of mathematics. This kind of dogmatism, of course, plagues any discipline and is not so easy to discern and sort out by the participants themselves. [W-3] is highly recommended reading.

Positional Topological Invariants

Much of Wilder's work can be said to center on placement problems and associated "positional topological invariants." These essentially mean properties of a space M in a space S which are independent of M's embedding in S.

For example, the uniform local connectedness of complementary domains of S^{n-1} in S^n is preserved by homeomorphs (in S^n) of S^{n-1}. These positional invariants often manifested themselves in the plane as thoroughly investigated set-theoretic concepts but which necessitate the introduction of homology to obtain generalizations to higher dimensions.

Consider the Jordan Curve theorem in the 2-sphere S^2. The complementary domains are exactly two uniformly locally connected subsets whose common boundary is the Jordan curve (S^1). Phrasing uniform local connectedness in homological terms is easy. Let D be one of the domains. For D to be uniformly locally connected means that given a finite open covering $\mathfrak{U}$ of S^2, there exists a finite open covering $\mathfrak{B}$ such that for each $V \in \mathfrak{B}$ there exists a $U \in \mathfrak{U}$ so that $\overline{H}_0(D \cap V) \to \overline{H}_0(D \cap U)$ is trivial, where $\overline{H}$ denotes augmented homology. Obviously, this depends only upon the frontier of a domain. This generalizes in a homological way to higher dimensions by defining i-ulc, $i \geqq 0$, to mean that the homomorphism, $\overline{H}_i(D \cap V) \to \overline{H}_i(D \cap U)$, induced by the inclusion is trivial. The domain is said to be ulc^r if it is i-ulc for all $0 \leq i \leq r$. Wilder showed in [W-1] and [W-2] that if M^{n-1} is a closed generalized manifold in S^n (or a spherelike generalized manifold), then both complementary domains are ulc^{n-1}. (The following example is instructive: Pinch a meridian on a standard torus in S^3 to a point. We have 2 complementary domains A and B. A fails to be 0-ulc but small 1-cycles bound, while B fails to have small 1-cycles bounding but is 0-ulc.)

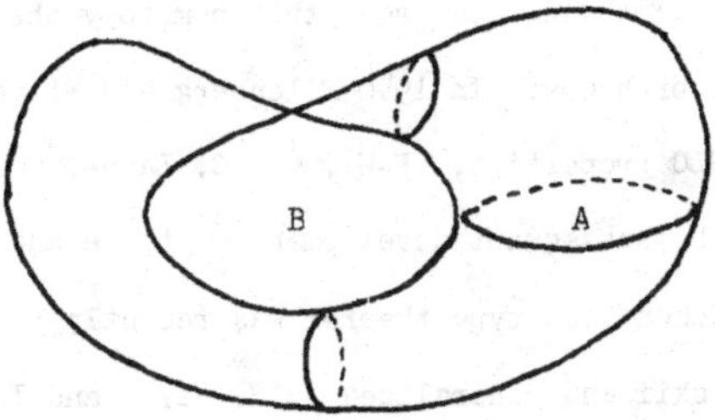

Conversely, he then showed that if a closed subset M of the sphere (or a spherelike generalized manifold) was the common boundary of at least 2 domains, one of which is ulc^{n-2}, then M is an orientable closed generalized $(n-1)$-manifold. Moreover, if S is an orientable n-dimensional generalized manifold such that $H_1(S)$ is trivial, and U is a ulc^{n-2} domain with connected boundary B, then B is an orientable $(n-1)$-dimensional generalized manifold. The result was a logical thorough extension and generalization of a theorem of Denjoy-R. L. Moore which stated that the boundary, if not a point, of a simply connected and uniformly locally connected subset of the 2-sphere was a circle, [M]. Wilder's extensive generalizations as found in Chapter X, especially sections 6 and 7, of his Colloquium publication, require the full power of the theory of generalized manifolds.

It may be appropriate at this time to mention that in the plane one can really go much further. Once a domain is bounded by a circle, the Schoenflies theorem tells us that the closure of the domain is a 2-cell. Alexander's example, as mentioned earlier, prevents a generalization of the Schonflies theorem to S^3 without strong additional hypotheses. The complementary domains are ulc^1 but they fail to be uniformly locally simply connected, 1-ULC. Of course, this was how Alexander recognized that a complementary domain was "bad." However, in the early 1930's homotopy theory hardly existed and researchers naturally thought in terms of homology. Nevertheless, in 1933 Wilder had shown that if U was an open subset of S^n and $M = \overline{U} - U$ was deformation free into U, (that is there existed a map $H: M \times I \to \overline{U}$, with $H(m,0) = m$, and $H(m,t) \in U$, for all $t > 0$) then M was a generalized $(n-1)$-manifold, [W-8]. With Hopf's and Hurewicz's work came recognition of the crucial role that homotopy theory would play in the solution of placement problems. In 1940 Eilenberg and Wilder published a fundamental paper on the ULC properties, [E-W]. R. C. Lacher has recently given an excellent survey of the subsequent development of these and related ideas in [La]. A special case of a Schonflies type theorem was recently proved independently by C. Seebech and Černavskiǐ and generalized by S. Ferry and T. Chapman: If M^{n-1} is a topological manifold embedded in an n-manifold N so that a complementary domain U is 1-ULC then its closure is an n-manifold with boundary. This theorem traces its roots to R. H. Bing's seminal paper [Bi].

"Topology of Manifolds"

In 1942 Wilder delivered the American Mathematical Society Colloquium Lectures. World War II intervened and consequently it was not until 1949 that "Topology of Manifolds" was published by the Society as volume 32 of its Colloquium Series. In the first portions of this 400 page book Wilder presents much of the topology of the plane that will be generalized to higher dimensions in the later portions of the book. Čech homology and cohomology theory and the lc^n and $colc^n$ properties are developed. At that time of writing and research, exact sequences and other functorial notions, which were to alter the point of view of topology, had not been introduced. Consequently, the reader will not find exact sequences, diagrams and sheaves explicitly mentioned in the text. Wilder by that time had settled upon a modification (deletion of a superfluous lc^n axiom) of E. Begle's definition of generalized manifold as well as Begle's proof of the duality theorems. This definition is essentially equivalent to the one most popular today. The book contained much of Wilder's previously unpublished research and many generalizations of his previous research.

New methods for obtaining Poincaré duality

About the time of the book's publication Leray introduced his ideas of sheaves and spectral sequences. H. Cartan combined these ideas and obtained proofs for Poincaré and Alexander Duality for manifolds using Leray's notion of "uniqueness" of a cohomology theory on a single space. Borel and Serre soon used spectral sequence arguments to reprove the Smith theorems. It was P. Conner, familiar with generalized manifold theory, the Smith theory and the "new" methods, who suggested that these theories were all interrelated and that sheaf theory and spectral sequences should be introduced into generalized manifold theory. A. Borel in [Bo-1] and I in [R-1] introduced homology theories equivalent to the Čech homology theory (over fields) which would enable Cartan's proof of Poincaré and Alexander duality to carry over to generalized manifolds. In 1961 Borel and Moore developed a homology theory which is exactly right for generalized manifolds over fairly arbitrary coefficients, [Bo-Mo]. This Borel-Moore homology theory is a modern reincarnation of the Steenrod "regular homology theory," [St]. P. Conner also suggested to me that it should be possible to show that if $M = A \times B$ was a generalized manifold then A and B would also have to be generalized manifolds. I announced a proof in 1958 which later appeared for the case of generalized manifolds with boundary, [R-2] (see also C. T. Yang [Y-1] and A. Borel [B-2; chap 1]).

Transformation Groups

Part of the reason for this flourish of activity in generalized manifolds was the proof by Conner and Floyd [C-F] that the so called Smith-manifolds were the same thing as generalized manifolds, [Sm]. It is strange that generalized manifolds and Smith-manifolds had co-existed independently, yet side by side, for such a long time. Smith had proved that the fixed point set of a prime power periodic map on a manifold would be a Smith-manifold. For a long time no really exotic fixed point sets of periodic maps on manifolds were known. Bing's many startling examples soon changed that. It was therefore absolutely necessary to consider generalized manifolds if a thorough treatment of topological transformation groups on classical manifolds was to be obtained.

The other crucial place where generalized manifolds fit into the topological theory of transformation groups arises from the theory of slices. That is if G is a compact Lie group acting on a (generalized) manifold M, then at each $x \in M$, there exists a set S_x such that an open invariant tubular neighborhood of the orbit $G(x)$ is homeomorphic to $S_x \times_{G_x} G$. This is a fiber bundle over the submanifold $G(x)$ with fiber, the slice S_x. The orbit space near x is just the quotient space S_x/G_x, where G_x is the isotropy group at x. Therefore, in the study of such group actions on generalized manifolds one may use inductive arguments since, in general, S_x will be a generalized manifold of dimension less than dimension M and G_x will be a proper closed subgroup of G. S_x is a generalized manifold because it is, of course, a factor of a generalized manifold. It is the well-known failure of factors of classical manifolds to retain their locally Euclidean character that makes the consideration of generalized manifolds necessary. An excellent and comprehensive treatment of topological transformation groups from the point of view of generalized manifolds can be found in [B-2].

Monotone Mapping Theorem

In 1956 Wilder proved his monotome mapping theorem [W-5] and [W-6]. Simply stated it says that if $f : M \to Y$ is a surjective mapping of an orientable generalized manifold onto a Hausdorff space Y, with $f^{-1}(y)$ a cyclic for all y, then Y is an orientable generalized manifold and f_* is a homology isomorphism. This is a far reaching generalization of R. L. Moore's monotone mapping theorem which stated that if Y was an upper semi-continuous decomposition of the 2-sphere S^2 by continua such that each non-degenerate element failed to separate S^2, then Y was homeomorphic to the 2-sphere. In Wilder's seminar there was a graduate student, Stephen Smale. Smale adapted some of Wilder's homological arguments to homotopy groups and obtained his second published paper a Vietoris type of mapping theorem for homotopy groups, [Sma]. (His first paper also arose out of discussions in this seminar.)

Michigan Topology Conference, 1966.

In 1966 Professor Wilder retired from active teaching at the University of Michigan and a Conference in Topology was organized in honor of his contributions to the mathematical community. Many very interesting papers were given at this conference and the second and third numbers of volume 14 (1967) of the Michigan Mathematical Journal are devoted to the proceedings. At that time several problems concerning generalized manifolds were discussed. Among these were [R-3]

(i) The Local Orientability Problem,

(ii) Triangulated Generalized Manifolds

(iii) The Relationship between the Hilbert-Smith Conjecture and Generalized Manifolds.

In the last 10 years remarkable progress has been made on the first 2 problems. The third remains as enigmatic as ever. I wish to discuss these now.

(i) <u>The Local Orientability Problem</u>

So far, I have avoided a definition of generalized manifold. If one traces the plethora of definitions one finds that at each new stage the definition becomes simpler and yet still incorporates the spaces satisfying the former definitions. Begle's definition, [Be-1] and [Be-2], as modified by Wilder, was particularly attractive except for the use of covering dimension and the notion of local orientability. Covering dimension had already slipped out of the picture by 1958. [It was replaced by finite cohomological dimension $(\dim_L X < \infty)$. The earlier problems of characterizing k-dimensional subsets of a generalized n-manifold were quite suitably answered in terms of the cohomological dimension and local linking instead of covering dimension.] Yet, the local orientability was still a "bug-a-boo." Let us now state the definition of a generalized manifold which takes into account G. Bredon's solution [Bre-2] of the local orientability problem. (Without Bredon's solution the definition would have to be a little more complicated.)

A locally compact Hausdorff space X is a generalized n-manifold over a principal ideal domain L if

1. $\dim_L X < \infty$

2. $H_j(X, X - x; L) = \begin{cases} 0, & j \neq n \\ L, & j = n \end{cases}$

3. X is locally connected in the sense of homology over L (lc_L^∞).

It is appropriate to take the Borel-Moore homology theory with coefficients in L [Bo-Mo]. When L is a field this coincides with Čech homology. If one chooses to use the singular homology theory instead, then one obtains what I called, in my thesis, singular homology n-manifolds. These are necessarily generalized n-manifolds and most of the usual theory works for singular homology manifolds too. If X is locally connected in the singular sense over L ${}^{s}\ell c_L^\infty$

it really doesn't matter which homology theory one uses, [Lee].

Condition 3 is a consequence of conditions 1 and 2 when L is a field and X is first countable. See Whitehead [Wh], Wilder [W-7] and Bredon [Bre-2].

The assignment to each open set U in X the L-module $H_n(X, X-\bar{U}; L)$ and the homomorphisms $j_* : H_n(X, X-\bar{U}; L) \to H_n(X, X-x; L)$, for each $x \in U$, yields a sheaf $\mathcal{O}$ on X, called the orientation sheaf. For almost all applications of the theory it was necessary to know that $\mathcal{O}$ was locally constant. Usually this could be deduced from the rest of the problem. Wilder, in equivalent terms, had asked, [W-4, p. 382], if $\mathcal{O}$ was locally constant. In 1969 Bredon in [Bre-1] showed under conditions 1 and 2 with only 0-lc_L from 3 that the orientation sheaf $\mathcal{O}$ is locally constant and settled this long standing problem.

The general form of Poincaré duality that holds for a generalized n-manifold is as follows: $H^p_\Phi(X; \mathcal{O}\otimes S) \xrightarrow{\cong} H^\Phi_{n-p}(X; S)$, where S denotes a coefficient sheaf of L-modules and Φ is any paracompactifying family. (Bredon had to use this most general form in his argument.) The duality is induced by a cap product and is natural with respect to inclusion maps and boundary-coboundary homomorphisms. See [Bo-Mo]. All standard types of duality, Poincaré, Alexander and Lefschetz are a form of this result and can essentially be deduced with only slight modifications from it.

A totally serious definition of generalized manifolds can be given as the class of spaces for which Poincaré duality holds both locally and globally.

(ii) Triangulated Generalized Manifolds

Ten years ago I observed that a genuine representable bordism theory may be obtained by replacing smooth manifolds by triangulated generalized manifolds, [R-3]. A main problem was to compute the stable homotopy groups of the associated spectrum. Analogues of the tangent bundle and normal bundle of a manifold are easy to construct. Certain types of characteristic classes had already been constructed, in particular the combinatorial Pontrjagin classes of Thom and Milnor. Martin and Maunder developed a very nice bundle theory modelled on block bundles, called homology cobordism bundles, [M-M], so that the analogue of Thom's bordism groups become the homotopy groups of a suitable Thom space. Maunder developed rational Pontrjagin classes for homology cobordism bundles in such a way that the familiar properties for vector bundles were preserved. Various types of general position were also developed by them. This led to the interesting work of Edmonds and Stern [E-S] and that of Galewski and Stern [G-S] and Matumoto [Ma] on the triangulability of topological manifolds. Some of this work has been reported on in this conference.

The most striking property exhibited by triangulated generalized manifolds M is:

$M \times R^1$ _is a topological manifold._

Of course, here the coefficients L must be taken to be the integers. This remarkable result is due to R. Edwards [E] and J. Cannon, [Ca]. It was long recognized that the singularities of a triangulated homology manifold start to arise from the links of the $(n-4)$-simplexes. But what is amazing is that all topological singularities are of codimension n (located at the vertices) and that they disappear in the product with a line.

Cannon in his work on this problem has conjectured:

Conjecture 1. If the ANR X is a generalized n-manifold (over Z), then $X \times R^1$ is a topological manifold.

Conjecture 2. If the ANR X is a generalized n-manifold over Z such that any 2 mappings of the 2-disk into X may be ϵ-homotoped off each other (the disjoint 2-disk property), then X is a topological manifold.

Along these lines Edwards has obtained:

A finite dimensional cell-like decomposition of a topological manifold which has the disjoint 2-disk property is a topological manifold.

Observe that this implies that any finite-dimensional cell-like decomposition of a manifold is at most a codimension two manifold factor. J. Bryant, J. Cannon and R. C. Lacher have obtained the following from still more general results:

If X is a finite dimensional ANR and a generalized n-manifold which fails to be locally Euclidean on a set of at most $\dim X/2$, then X is a manifold *if* (*and only if*) *it has the disjoint 2-disk property.*

Cell-like decompositions of a manifold, even though they are manifold factors, can be rather bizarre. By embedding countably many "distinct" wild arcs densely in any 3-manifold M, Kwun [K-2], has observed that the resulting generalized 3-manifold M^* obtained by collapsing each arc to a point has the property that each self-homeomorphism is the identity, (yet $M^* \times R^1$ is homeomorphic to $M \times R^1$). It seems that for some problems researchers are pushing in the right directions by attempting to "resolve" generalized manifolds rather than trying to work directly with them.

Another direction for investigation would be a more comprehensive study of triangulated generalized manifolds over coefficients other than the integers. One reason why this is interesting is that if we begin with an oriented (generalized) manifold M and let a finite group G of orientation preserving homeomorphisms act, then M/G is an oriented generalized n-manifold over any field whose characteristic is prime to the order of G, [R-5]. In particular, if G acts smoothly or simplicially M/G would be triangulated. For example, a finite group of algebraic automorphisms of a non-singular algebraic variety would give M/G, a singular algebraic variety, the structure of a triangulated generalized manifold over fields whose coefficients are prime to the orders of the stability groups of G. One has rational Pontrjagin classes and much of the theory of Martin-Maunder as well as the Galewski-Stern theory carries over.

(iii) The Generalized Hilbert-Smith Conjecture

This states that any compact group of homeomorphisms of a topological manifold must be a Lie group. The same conjecture is also unknown for generalized manifolds.

The conjecture holds for manifolds of dimension less than 3, [M-Z, p. 249] and for smooth manifolds, if G is a group of diffeomorphisms [M-Z, p. 208] (in fact, Lipschitz homeomorphisms, [H]). Other special conditions are known which restrict G. For example, if all orbits are locally connected, then G is a Lie group, [M-Z, p. 244].

How does this conjecture tie in with generalized manifolds? As mentioned above, M^m/G is an m-generalized manifold over Q if a finite group G acts preserving orientation. This result still holds for any 0-dimensional compact group, [R-5]. The structure theory of compact groups shows that the conjecture holds if and only if no p-adic group A_p can act on the manifold. ($A_p = \text{Inv}\lim(Z_p \leftarrow Z_{p^2} \leftarrow Z_{p^3} \leftarrow \ldots)$).

One way to investigate the conjecture is to assume that one does have an effective A_p action on a manifold. By examining the consequences one would hope to either arrive at a contradiction or be led by the necessary consequences to a construction of such an example. Since no recent progress has been made with this approach, I will try to re-interest people in the problems I proposed in [R-3]. Let us assume that we have an effective action of A_p on a manifold M, $A_p \times M \to M$. Let Σ_p denote the p-adic solenoid.

$$(\Sigma_p = \text{Inv}\lim(S^1 \xleftarrow{\text{degree } p} S^1 \xleftarrow{\text{degree } p^2} S^1 \xleftarrow{p^3} S^1 \leftarrow \ldots)).$$

Form the product $\Sigma_p \times M$ and let A_p act diagonally. Then this free action of A_p will commute with that of Σ_p which acts on $\Sigma_p \times M$ by translating on the first factor. We examine the commutative diagram of equivariant, orbit and fiber maps

$$\begin{array}{ccccc}
(\Sigma_p, A_p) & \xleftarrow{/M} & (\Sigma_p, \Sigma_p \times M, A_p) & \xrightarrow{/\Sigma_p} & (M, A_p) \\
\downarrow /A_p & & \downarrow /A_p & & \downarrow /A_p \\
\Sigma_p/A_p = S^1 & \xleftarrow[\nu]{/M} & (\Sigma_p, \Sigma_p \times_{A_p} M) & \xrightarrow[\mu]{/\Sigma_p} & M/A_p .
\end{array}$$

The map ν is a fibering over the circle with fiber the manifold M^m and structure group A_p. Therefore, $\Sigma_p \times_{A_p} M$ is an $(m+1)$-manifold, say N. The p-adic solenoid acts on N^{m+1}. The orbit space of this action is M/A_p, the orbit space of the A_p action on M^m. No generality is lost in assuming both that M is oriented and that the action of A_p preserves orientation. Hence, by [R-5],

M/A_p is an m-generalized manifold over every field whose characteristic is not p. If the action of Σ_p is effective then we have, [Y-1] and [B-R-W],:

<u>PROPOSITION</u>: $\dim_L(M/A_p) = \begin{cases} m, \text{ if } L \text{ is a field, } \chi(L) \neq p \\ m+1, \text{ if } L \text{ is a field, } \chi(L) = p \\ m+2, \text{ if } L \cong Z. \end{cases}$

Let us examine this for the special case when the action of A_p is <u>free</u>. Then the action of $(\Sigma_p, \Sigma_p \times_{A_p} M)$ is also free and $\mu^{-1}(n^*) = \Sigma_p$, for each $n^* \in N/\Sigma_p = M/A_p$. If we use $L = Z/pZ$ coefficients, the mapping μ is a proper <u>acyclic</u> (over L) mapping. By Wilder's monotone mapping theorem, [W-5], $M/A_p = N^{m+1}/\Sigma_p$ will necessarily be an oriented $(m+1)$-generalized manifold. Thus,

M/A_p <u>is an orientable</u>

1. $(m+1)$-<u>generalized manifold over</u> Z/pZ, <u>and an</u>
2. m-<u>generalized manifold over any field</u> L <u>whose characteristic</u> $\neq p$, <u>and</u>
3. <u>yet</u> $\dim_Z M/A_p = m + 2$.

It should be mentioned that M/A_p can not be an integral generalized manifold. In fact, <u>it will not be</u> clc <u>over</u> Z, (the $m+2$-local cohomology groups over Z are weird).

Of course, a positive response to the following would yield a proof of the Hilbert-Smith Conjecture.

<u>Question</u>. Let p be a fixed prime. Let M be a generalized m-<u>manifold for each prime field whose characteristic</u> $\neq p$. <u>If</u> $\dim_Z M < \infty$, <u>then is</u> $\dim_{Z/pZ} M = m$?

Nothing negative is known even where M is only assumed to be a generalized m-manifold over <u>some</u> field whose characteristic $\neq p$. However, a positive answer to the following is likely to be correct. (Versions with somewhat stronger hypothesis are known to hold.) Unfortunately, this is not relevant to the Hilbert-Smith Conjecture.

Question. Let M be clc over $\mathbb{Z}$ with $\dim_{\mathbb{Z}} M < \infty$. If M is an orientable generalized m_p-manifold over each prime field $\mathbb{Z}/p\mathbb{Z}$ and $\mathbb{Q}$, then is M a generalized m-manifold over $\mathbb{Z}$? (If so, $m = m_p$ for each prime p and $\mathbb{Q}$.)

Further discussion of the Hilbert-Smith Conjecture can be found in [R-4].

Conclusion.

The subject of generalized manifolds arose naturally from many directions. This diversity of its sources lent much to its vitality and usefulness to various parts of mathematics. Many thanks are due to R. L. Wilder whose pioneering work in this area began some 50 years ago and continued for over 40 years. Today the subject has Professor Wilder's indelible mark on it. It is mature, active and with important problems still remaining to be solved.

PUBLICATIONS OF RAYMOND L. WILDER

1924

On the dispersion sets of connected point sets. Fund. Math. 6, 214-228.

On a certain type of connected set which cuts the plane. Proc. Inter. Math. Cong. Toronto, 423-437.

1925

Concerning continuous curves. Fund. Math. 7, pp. 340-377. (Dissertation).

A theorem on continua. Fund. Math. 7, 311-313.

A property which characterizes continuous curves. Proc. National Acad. of Sci. 11, 725-728.

1926

A theorem on connected point sets which are connected in kleinen. Bull. Amer. Math. Soc. 32, 338-340.

1927

A connected and regular point set which has no subcontinuum. Trans. Amer. Math. Soc. 29, 332-340.

The non-existence of a certain type of regular point set. Bull. Amer. Math. Soc. 3, 439-446.

A point set which has no true quasi-components, and which becomes connected upon the addition of a single point. Bull. Amer. Math. Soc. 33, 423-427.

1928

On connected and regular point sets. Bull. Amer. Math. Soc. 34, 649-655.

Concerning R. L. Moore's axiom Σ_1 for plane analysis situs. Bull. Amer. Math. Soc. 34, 752-760.

A characterization of continuous curves by a property of their open subsets. Fund. Math. 11, 127-131.

1929

Review of Fraenkel on Grundlegung der Mengenlehre. Bull. Amer. Math. Soc. 35, 405-406.

Concerning zero-dimensional sets in euclidean space. Trans. Amer. Math. Soc. 31, 405-406.

Characterizations of continuous curves that are perfectly continuous. Proc. Nat. Acad. Sci. 13, 614-621.

Analysis situs (position analysis). Encyclopedia Brittannica. 14th Ed. 1, 865-867.

Point sets. Encyclopedia Brittannica. 14th Ed. 18, 117-118.

1930

Concerning perfect continuous curves. Proc. Nat. Acad. Sci. 16, 233-240.

A converse of the Jordan-Brouwer separation theorem in three dimensions. Trans. Amer. Math. Soc. 32, 632-657.

1931

Concerning simple continuous curves and related poincs sets. Amer. J. Math. 53, 39-55.

Extension of a theorem of Mazurkiewiez. Bull. Amer. Math. Soc. 37, 287-293.

1932

A plane, arewiss connected and connected in kleinen point set which is not strongly connected in kleinen. Bull. Amer. Math. Soc. 38, 531-532.

Point sets in three higher dimensions and their investigation by means of a unified analysis situs. Bull. Amer. Math. Soc. 38, 649-692.

On the imbedding of subsets of a metric space in Jordan continua. Fund. Math. 19, 45-64.

1933

On the linking of Jordan continua in E_n by (n-2)-cycles. Ann. of Math. 34, 441-449.

On the properties of domains and their boundaries in E_n. Math. Annalen. 109, 272-306.

Concerning a problem of K. Borsuk. Fund. Math. 21, 156-167.

Review of S. Lefschetz, Topology. Amer. Math. Monthly 40, 232-233.

1934

Concerning irreducibly connected sets and irreducible regular connexes. Amer. J. Math. 56, 547-557.

Generalized closed manifold in n-space. Annals of Math. 35, 876-903.

1935

On free subsets of E_n. Fund. Math. 25, 200-208.

On locally connected spaces. Duke Math. J. 1, 543-555.

1936

A characterization of manifold boundaries in E_n dependent only on lower dimensional connectivities of the complement. Bull. Amer. Math. Soc. 42, 436-441.

The strong symmetrical cut sets of closed Euclidean n-space. Fund. Math. 27, 136-139.

1937

Some unsolved problems of topology. Amer. Math. Monthly 44, 61-70.

1938

Sets which satisfy certain avoidability conditions. Casopis Pro Pestovani Matematiky A Fysiky 67, 185-198.

1939

The sphere in topology. Semicentennial Addresses of the Amer. Math. Soc.

Review of M. H. A. Newman, Plane Topology. Science 90, 354-355.

Property S_n. Amer. J. Math. 61, 823-832.

1941

Uniform local connectedness. In Lectures in Topology. Univ. of Mich. Press, Ann Arbor. 1941, 29-41.

Decompositions of compact metric spaces. Amer. J. Math. 63, 691-697.

1942

Uniform local connectedness and contractibility (with S. Eilenberg). Amer. J. of Math. 64, 613-622.

1944

The nature of mathematical proof. Amer. Math. Monthly 51, 309-323.

1949

Topology of manifolds. Amer. Math. Soc. Colloquium Publ., Vol. 32, ix + 402.

1950

The cultural basis of mathematics. Proc. Inter. Cong. of Mathematicians, Cambridge, I, 258-271.

A generalization of a theorem of Pontrjagin. Ibid., 530-531.

1952

Introduction to the foundations of mathematics. New York: Wiley and Sons, xiv + 305 pp.

1953

The origin and growth of mathematical concepts. Bull. Amer. Math. Soc. 59, 423-448.

1954

On certain inequalities relating the Betti numbers of a manifold and its subsets (with J. P. Roth). Proc. Mat. Acad. Sci. 40, 207-209.

Review of H. Hasse, Mathematik als Wissenchaft, Kunst and Macht. Bull. Amer. Math. Soc. 60, 181-182.

Review of W. Sierpinski. General Topology. Scripta Mathematica 20, 84-86.

A type of connectivity. Proc. Inter. Cong. Mathematicians. Amsterdam 2, 264 (Abstract).

1955

Concerning a problem of Alexandroff. Mich. Math. J. 3, 181-185.

1956

Review of J. L. Kelley. General Topology. Scripta Mathematica 22, 255-256.

1957

Some consequences of a method of proof of J. H. C. Whitehead. Mich. Math. J. 4, 27-31.

Some mapping theorems with applications to non-locally connected spaces. Algebraic Geometry and Topology. A Symposium in honor of S. Lefschetz. Princeton University Press, 377-388.

Monotone mappings of manifolds. Pacific J. of Math. 7, 1519-1528.

1958

Monotone mappings of manifolds II. Mich. Math. J. 5, 19-23.

Local orientability. Colloquium Mathematicum 6, 79-93.

Review of R. L. Goodstein. Mathematical Logic. Math. Rev. 19, 1-2.

1959

The existence of certain types of manifolds (with M. L. Curtis). Transactions of the AMS 91, 152-160.

Axiomatics and the development of creative talent. The Axiomatic Method. Ed. by L. Henkin, P. Suppes, and A. Tarski. Amsterdam, 474-488.

The nature of modern mathematics (Russel Lecture for 1959). Mich. Alumnus Quarterly Review 65, 302-312.

1960

Mathematics: A cultural phenomenon. In "Essays in the Science of Culture". Ed. by G. E. Dole and R. L. Carneiro, New York, 471-485.

A certain class of topological properties. Bull. Amer. Math. Soc. 66, 205-239.

1961

Extension of local and medial properties to compactifications with an application to Čech manifolds. Czech. Math. J. 11, 306-318.

A converse of a theorem of R. H. Bing and its generalization. Fund. Math. 50, 119-122.

Material and method, in "Undergraduate Research in Mathematics", K. O. May and S. Schuster ed., Northfield, Minn., 9-27.

1962

Freeness in n-space, in "Topology of 3-Manifolds and Related Topics", M. R. Fort, ed., New York, Prentice-Hall, 106-109.

Partially free subsets of Euclidean n-space. Mich. Math. J. 9, 97-107.

Topology: Its nature and significance. The Math. Teacher, 55, 462-475.

1963

Axiomatization. The Harper Encyclopedia of Science. 1963, p. 128.

Axiom of Choice. The Harper Encyclopedia of Science. 1963, p. 128.

Topology. The Harper Encyclopedia of Science, 1193-1194.

1964

General Topology. Encyclopedia Britannica 22, 298-301.

Point Sets, Encyclopedia Britannica 18, p. 187.

1965

A problem of Bing, Proc. Nat. Acad. Sci. 54, 683-687.

1966

An elementary property of closed coverings of manifolds, Mich. Math. J. 13, 49-55.

1967

The role of the axiomatic method, Amer. Math. Monthly 74, 115-127.

The nature and role of research in mathematics, in "Research: Definitions and Reflections (Essays on the occasion of the Univ. of Michigan's Sesquicentennial)", Univer. of Mich., Ann Arbor, 96-109.

The role of intuition, Science 156, 605-610.

1968

Addition and reduction theorems for medial properties, Trans. Amer. Math. Soc. 130, 131-140.

Evolution of Mathematical Concepts, Wiley, New York.

1969

Mathematics' Biotic Origins, Medical Opinion and Review 5, 124-135.

Trends and social implications of research, Bull. Amer. Math. Soc. 75, 891-906.

Development of modern mathematics, in "Historical Topics for the Mathematics Classroom", 31st Yearbook, Nat'l Council of Teachers of Math., Washington, D.C., 460-476.

The nature of research in Mathematics, in "The Spirit and Uses of the Mathematical Sciences", McGraw-Hill, New York, 31-47.

1970

The beginning teacher of college mathematics, in "Effective College Teaching", Amer. Council on Educ., Washington, D.C., 94-103.

Historical background of innovations in mathematics curricula, in "Mathematics Education", 69th Yearbook of Nat'l Soc. for the Study of Educ., Part I, Univ. of Chic. Pr., 7-22.

The beginning teacher of college mathematics, CUPM Newsletter, December, No. 6.

1972

The Nature of Modern Mathematics, in "Learning and the Nature of Mathematics, ed. W. E. Lamon, Chicago, Science Research Associates, Inc., 35-48.

History in the Mathematics Curriculum: Its Status, Quality and Function, Amer. Math. Monthly 79, 479-495. [This paper received a Lester R. Ford Award from the Mathematical Association of America, Jan. 1973.] [Reprinted as La Historia en el programa de Matematicas: su estado calidad y funcion, Boletin de Matematicas 6 (1972), 23-58.]

1973

Mathematical Rigor, Relativity of Standards of, in "Dictionary of the History of Ideas", New York, Chas. Scribner's Sons, Vol. 3, 170-177.

Recollections and Reflections, Math. Mag. 46, 177-182.

Mathematics and Its Relations to Other Disciplines, The Math. Teacher 66, 679-685.

1974

Hereditary Stress as a Cultural Force in Mathematics, Historia Mathematica 1, 29-46.

1975

Review of Claudia Zaslavsky, "Africa Counts", Boston, Prindle, Weber and Schmidt, 1973, in Historia Mathematica 2, 207-210.

1976

Commentary on 4 papers in Norbent Wiener, Collected Works, Cambridge, Mass., MIT Press, Vol. 1.

Robert Lee Moore, 1882-1974, Bull. Amer. Math. Soc. 82, 417-427.

References

An extensive list of references can be found in [W-4]. Also, Professor Wilder, in the reprinted edition of 1963, p. xiv, updated numerous questions and problems he had posed in the first edition. I have included a few relevant references that have not been used explicitly in the text.

[A-1] Alexander, J. W., A proof and extension of the Jordan-Browner separation theorem, Trans. Amer. Math. Soc., vol. 23 (1922), 333-349.

[A-2] ________, An example of a simply connected surface bounding a region which is not simply connected, Proc. Nat. Acad. of Sciences, vol. 10 (1924), 8-10.

[A-P] Alexandroff, P., and Pontrjagin, L, Les variétés â n-dimensions généralisées, Comptes Rendus de l'Académie des Sciences, Paris, vol. 202 (1936), 1327-29.

[Be-1] Begle, E., Locally connected spaces and generalized manifolds, Amer. Jour. Math., 64 (1940), 553-574.

[BE-2] ________, Duality theorems for generalized manifolds, Amer. Jour. Math., 67 (1945), 59-70.

[Bi] Bing, R. H., A surface is tame if its complement is 1-ULC, Trans. Amer. Math. Soc., 101 (1961), 294-305.

[Bo-1] Borel, A., The Poincaré duality in generalized manifolds, Mich. Math. Jour., 4 (1957), 227-239.

[Bo-2] Borel, A., et al, Seminar in Transformation Groups, Annals of Math. Study, 46, Princeton, 1960.

[Bo-Mo] Borel, A. and Moore, J., Homology theory for locally compact spaces, Mich. Math. Jour., 7 (1960), 137-160.

[Br] Brahana, T. R., A theorem about local Betti groups, Mich. Math. Jour., 4 (1957), 33-38.

[Bre-1] Bredon, G. E., Orientation in generalized manifolds and applications to the theory of transformation groups, Mich. Math. Jour., 7 (1960), 35-64.

[Bre-2] ______, Wilder manifolds are locally orientable, Proc. Nat. Acad. Sci., U. S., 63 (1969), 1079-1081.

[Bre-3] ______, Generalized manifolds, revisited, "Topology of Manifolds," Markham, 1970, 461-469.

[B-R-W] Bredon, G. E., Raymond, F. and Williams, R. F., p-adic groups of transformations, Trans. Amer. Math. Soc. 99 (1961), 488-498.

[Ca] Cannon, J. The double suspension of 3 dimensional homology spheres is S^5. [to appear].

[C-1] Čech, E., Théorie générale de l'homologie dans un espace quelconque, Fund. Math., vol. 19 (1932), 149-183.

[C-2] ______, Sur la dimension des espaces parfaitement normaux, Bulletin international de l'Académie des Sciences de Bohême, 1932, pp. 1-25.

[C-3] ______, On general manifolds, Proceedings of the National Academy of Sciences, vol. 22 (1936) 110-111.

[C-4] ______, On pseudomanifolds (mimeographed), Princeton, N. J., The Institute for Advanced Study, 1935.

[C-5] ______, Sur la décomposition d'une pseudovariété par un sousensemble fermé, Comptes Rendus de l'Académie des Sciences, vol. 198 (1934), 1342-1345.

[Co] Conner, P. E., On the action of the circle group, Mich. Math. Jour., 4 (1957), 241-248.

[C-D] Conner, P. E., and Dyer, E., On singular fiberings of spheres, Mich. Math. Jour., 6 (1959), 303-312.

[C-F] Conner, P. E. and Floyd, E. E., A characterization of generalized manifolds, Mich. Math. Jour., 6 (1959), 33-43.

[Cu] Curtis, M. L., Smoothness conditions on continua in Euclidean space, Mich. Math. Jour., 14 (1967), 277-282.

[C-W] Curtis, M. L. and Wilder, R. L., The existence of certain types of manifolds, Trans. Amer. Math. Soc., 91 (1959), 152-160.

[E] Edwards, R. D., On the double suspension of certain homology spheres [to appear].

[E-S] Edmonds, A. and Stern, R., Resolutions of homology manifolds: A classification theorem, J. London Math. Soc. (2), 11 (1975), 474-480.

[E-W] Eilenberg, S. and Wilder, R. L., Uniform local connectedness and contractibility, Amer. Jour. Math., 64 (1942), 613-622.

[G-S] Galewski, D. and Stern, R., Classification of simplicial triangulations of topological manifolds, Bull. Amer. Math. Soc., 82 (1976), 916-918.

[H] Hahn, F. J., On the action of a locally compact group on E_n, Pac. Jour. Math., 11 (1961), 221-223.

[J] Jones, Burton F., Wilder on connectedness, these proceedings.

[Ka] Kato, M., Partial Poincaré duality for k-regular spaces and complex algebraic sets, Topology, 16 (1977), 33-50.

[K-1] Kwun, K. W., Examples of generalized-manifold approaches to topological manifolds, Mich. Math. Jour., 14 (1967), 225-230.

[K-2] ______, A generalized manifold, Mich. Math. Jour., 6 (1959), 299-302.

[La] Lacher, R. C., Cell-like mappings and their generalizations, Bull. Amer. Math. Soc., 83 (1977), 495-552.

[Lee] Lee, C. N., The regular convergence theorem, Mich. Math. Jour., 14 (1967), 207-218.

[Le] Lefschetz, S., On generalized manifolds, Amer. J. Math., vol. 55 (1933), 469-504.

[M-M] Martin, N. and Maunder, C., Homology cobordism bundles, Topology, 10 (1971), 93-110.

[Ma] Matumoto, T., Variétés simpliciales d'homologie et variétés topologiques métrisable, thèse, Univ. de Paris-Sud, 1976.

[M] Moore, R. L., A characterization of Jordan regions by properties having no reference to their boundaries, Proc. Nat. Acad. Sci., 4 (1918) 364-370.

[M-Z] Montgomery, D. and Zippin, L., Topological Transformation Groups, Interscience, New York, 1955.

[R-1] Raymond, F., Čech homology from a chain complex, Abstract 542-125, Notices Amer. Math. Soc., 5 (1958), 203.

[R-2] ______, Separation and union theorems for generalized manifolds with boundary, Mich. Math. Jour., 7 (1960), 7-21.

[R-3] ______, Two problems in the theory of generalized manifolds, Mich. Math. Jour., 14 (1967), 353-356.

[R-4] ______, Cohomological and dimension theoretical properties of orbit spaces of p-adic actions, Proc. Conf. Transf. Groups, New Orleans, 1967, Springer-Verlag, New York, 1968, 354-365.

[R-5] ______, The orbit space of totally disconnected groups of transformations on manifolds, Proc. Amer. Math. Soc., 12 (1961), 1-7.

[Sma] Smale, Stephen, A Vietoris mapping theorem for homotopy, Proc. Amer. Math. Soc., 8 (1957), 604-610.

[Sm] Smith, P. A., Transformations of finite period II, Annals of Math., (2), 40 (1932), 690-711.

[St] Steenrod, N. E., Regular cycles of compact metric spaces, Annals of Math., 41 (1940), 833-851.

[V.K.] Van Kampen, E. R., Die Kombinatorische Topologie und die Dualitätssütze, The Hague, 1929, Leyden thesis.

[Wh] Whitehead, J. H. C., Note on the condition n-colc, Mich. Math. Jour., 4 (1957), 25-26.

[W-1] Wilder, R. L., A converse of the Jordan-Brouwer separation theorem in three dimensions, Trans. Amer. Math. Soc., 32 (1930), 632-657.

[W-2] ______, Generalized closed manifolds in n-space, Annals of Math., 35 (1934), 876-903.

[W-3] ______, Point sets in three and higher dimensions and their investigation by means of a unified analyses situs, Bull. Amer. Math. Soc., 38 (1932), 649-692.

[W-4] ______, Topology of Manifolds, Amer. Math. Soc., Colloquium Publ. 32, 1949.

[W-5] ______, Monotone mappings of manifolds, Pac. Jour. Math., 7 (1957), 1519-1528.

[W-6] ______, Monotone mappings of manifolds II, Mich. Math. Jour., 5 (1958), 19-23.

[W-7] ______, Some consequences of a method of proof of J. H. C. Whitehead, Mich. Math. Jour., 4 (1957), 27-32.

[W-8] ______, Concerning a problem of K. Borsuk, Fund. Math., 21 (1933), 156-167.

[Y-1] Yang, C. T., Transformation groups on a homological manifold, Trans. Amer. Math. Soc., 87 (1958), 261-283.

[Y-2] ______, p-adic transformation groups, Mich. Math. Jour., 7 (1960), 201-218.

R.L. MOORE'S FIRST DOCTORAL STUDENT AT TEXAS

by Lucille Whyburn

The University of Texas at Austin

It seems right and proper that those who have only known R.L. Wilder as Research Professor of Mathematics at the University of Michigan, as President of the AMS and MAA, as recipient of the Distinguished Award for Service to Mathematics, and as a member of that exclusive group known as the National Academy of Sciences, should stop to think that he was once a struggling student and an untenured faculty member. Indeed he was R.L. Moore's first doctoral candidate at the University of Texas.

Recently while working through the Moore Archives at the University of Texas, I opened Wilder's folder and, atop all else, I found the following statement in Moore's own handwriting:

> "Of all the men who have received the Ph.D. degree at the University of Texas, Dr. Wilder is one of the very best. He has certainly demonstrated unusual capacity for productive scholarship and I consider him exceptionally well equipped to attack certain problems of fundamental interest."

Although one of Moore's students was to write back to him after leaving Texas, saying "All of us used to think that you could pick out a Topologist by the look in his eye at birth," Moore was cautiously reluctant to admit Ray into his course on "Foundations of Analysis Situs". He openly questioned why a person interested in actuarial work might think he would be interested in pure mathematics. He conjectured that Ray would neither be good at nor enjoy finding rigorous proofs. The battle was joined--skepticism against ability fortressed by determination. Ray persisted, saying he would like to try. Moore subjected Wilder to extensive questioning. Pleased and surprised by Ray's answer to the question "What is an axiom?", Moore finally granted him admission but ignored him at first in the class. However, when Wilder proved Theorem 15 [1], things began to change. Then when Ray found the solution of a problem that Moore and J.R. Kline had been working on, Moore invited him to write it up as a Ph.D. thesis, cut through red tape to push through the awarding of his Ph.D. degree that year, and became his lifelong friend. Thus as Moore's first Ph.D. at the University of Texas, Ray stands at the head of the line of Moore's distinguished students at this institution.

The following letters from the Moore Collection were written over 50 years ago and each one has a charm all its very own. The first one reveals Wilder's friendship with Moore. Ray, his wife Una, and their infant daughter Mary Jane had set out from Texas for Ohio in September, 1924. Shortly a letter came back to Austin, saying:

"Columbus, Ohio
September 28, 1924

Dear Dr. Moore,

Please excuse paper, pencil, and writing, as this is being done, under difficulties, on a book held in my lap.

We accomplished our journey without mishap, both Mary Jane and myself behaving well (not to imply, however, that Mrs. Wilder did not). Kuhn [2] met us at the station, and took us to a nice boarding house not far from the University .....

It is too early to say yet how I am going to fit in here. One minute I have grave doubts and the next I feel reassured. In the first place, Kuhn seems to favor the oath [3] - says it may be a nuisance, but when it's <u>needed</u> it's a good thing.

Well today I got at least two hopeful signs. I tackled Arnold on the oath question and he agreed with me. Also agreed with me on professional patriotism. Then Kuhn called this afternoon and began telling me of graduate students who might take the Analysis Situs if I gave it ... , told me of five, and I am going to see them Tuesday to get an idea of what they are like ...

So things are looking a little brighter now."

Note from the above excerpt of Wilder's letter that even in 1924 he was a liberal and objected to the loyalty oath that followed World War I.

The next letter shows that R.L. Wilder's enthusiasm for finding talent began early and that he was then, as he has remained, a very careful scholar.

"Columbus, Ohio
Feb. 6, 1925

Dear Dr. Moore:

I am teaching the "Foundations" this quarter. I have three graduate students (two men and one woman) and an undergraduate man visiting.... So far the three graduate students have done pretty well. One of the men looks as tho he is pretty good material... only he's one of these fellows that try to make anything difficult,... he gets there, tho. Also the woman shows up pretty well. None has proved Theorem 15 however. They seem greatly interested.... They told me this morning that the only trouble with the course was that it hurt their other courses.... They get working on a problem and can't drop it.... I hope it is true, stick-to-it-iveness helps! They have actually proposed and proved several little original theorems that aren't so bad, and we're keeping an 'Auxiliary' set of lettered theorems for such. Believe me, your method of teaching is the only successful one.

I want to ask you about a point in your paper "Concerning the prime parts of certain continua which separate the plane", (Proc. Nat. Acad. Sci. Vol. 10, 1924). I was reading your proof of Lemma 3, and got stuck on the fourth sentence from the end.... 'Thus D_x contains a point of E and therefore, since E is connected and contains no point of the boundary of D_x ....', the underlined part is what I refer to. I don't say it is wrong, for I am pretty weak on 'prime parts' and am just trying to get on a familiar basis with them. Then the following example oc-

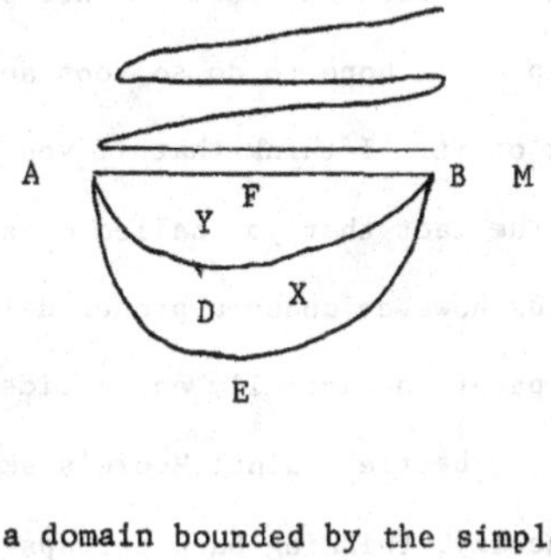

curred to me: Let D be a domain bounded by the simple closed curve ACBEA. The arc AFB being a condensation set of a sine curve is a prime part of M, isn't it?

Then it seems as though K will consist of all points of ACBEA - (A + B) , together with the prime part AFB . But if AFB = p , K - p is not connected, contrary to your lemma. In this example E does contain points of the boundary of D_x (H* = simple closed curve AFBCA) .

If I have made some mistake in this, I hope you'll set me to rights. I noticed you used this lemma only to prove Lemma 8, and thus you do not need such a strong theorem. I think it can be shown that B cannot contain uncountably many prime parts of M such that the omission of any one of them disconnects B - the property you used in proving Lemma 8.

I've started a bi-weekly club here, such as we had in Texas."

R.L. Moore replies to this letter:

"Feb. 23, 1925

Dear Dr. Wilder:

You are right about that point. I am glad you noticed it.... I soon found however after getting your letter, that I had not properly defined the outer boundary of D with respect to the prime parts of M . With a proper definition of this notion Lemma 3 is all right, and the argument as it stands will then be all right after a modification of the final sentence to conform to the changed definition in question.

I will enclose a reprint with these changes indicated.... I have been trying to get into shape a paper which contains a very general result in view of which I felt that Lemma 3 must be right unless some term was not properly defined. I have not yet finished writing it up but I hope to do so soon and I think I shall send you at least an approximate draft of it. I think that if you don't object, I will in this paper make reference to the fact that you called my attention to this error, and indicate that Lemma 3 holds however under a proper definition of outer boundary.

I am glad you read this paper so carefully and noticed that bad place. Thanks!"

Ray Wilder not only won the battle against Moore's skepticism but, after joining the University of Michigan faculty, thinking back perhaps to his own student days, he created his famous course "Foundations of Mathematics". It is interesting that he began this course in the early thirties with a class of approximately thirty students whose central interest was actuarial mathematics. Much later he was to write a textbook for such a course and in the preface he says, "The reason for instigating such a course was simply the conviction that it was not good to have teachers, actuaries, statisticians, and others who had specialized in undergraduate mathematics, and who were to base their life's work on mathematics, leave the university without some knowledge of modern mathematics and its foundations." Here was a scholar, thrilled by and active in the creation of modern mathematics, who was sharing with these students his way of thinking, who was able and willing to accept and use spontaneous ideas and illustrations, who was not afraid to introduce and discuss controversial points, who was showing the students how to think about mathematics, not just accept it as a dogmatic type of discipline.

This course served many purposes. It illustrated perfectly the leavening effect of a research scholar as a teacher. It attracted exceptional students from the social sciences and philosophy, offering them an opportunity to get a feeling for modern mathematics, and even enabling some of them to put those concepts to use in their own fields. Mathematics majors found ideas and methods of proof that would prove highly useful if they entered graduate school.

Thus we see R.L. Wilder developed into a teacher capable of teaching well at both the undergraduate and graduate level, became a leader in his department and university, and produced fundamental research in pure mathematics. Furthermore, he contributed to the education of scholars at large through his participation in mathematical forums such as this Symposium; and a mathematical conference is the most exacting class of all since it is an assembly of experts.

[1] This was Theorem 15 in R.L. Moore, Foundations of Plane Analysis Situs, _Trans. Amer. Math. Soc._ 17 (1916), 131-164.

[2] Harry Waldo Kuhn

[3] At that time, prospective employees of Ohio State University were required to take a broad oath of allegiance to both the U.S. and the State of Ohio, including a pledge to obey all their laws. (At that time, the controversial "Prohibition Laws" were in effect.)

A HOMOLOGICAL CHARACTERIZATION OF INNER AUTOMORPHISMS OF COMPACT LIE GROUPS

Robert F. Brown

University of California, Los Angeles

Let G be a compact, connected Lie group. Suppose that $h : G \to G$ is an inner automorphism, that is, there exists $a \in G$ such that $h(x) = axa^{-1}$ for all $x \in G$. Since G is pathwise-connected, there is a map $C : [0,1] \to G$ such that $C(0) = a$ and $C(1) = e$, the unit element of G. The map $H : G \times [0,1] \to G$ defined by $H(x,t) = C(t)x(C(t))^{-1}$ is a homotopy between h and the identity automorphism on G. Consequently, from the point of view of algebraic topology, an inner automorphism is indistinguishable from the identity because, for any homotopy-invariant functor F, $F(h) = F(\text{identity})$.

We are then led to ask, for what functor F is the converse true? That is, is there a homotopy-invariant functor F such that if $h : G \to G$ is an endomorphism with $F(h) = F(\text{identity})$, then h is an inner automorphism? The answer is given by the following characterization theorem.

THEOREM. <u>Let</u> $h : G \to G$ <u>be an endomorphism of a compact, connected Lie group</u>, <u>then</u> h <u>is an inner automorphism if and only if the induced endomorphism of real homology</u>

$$h_* : H_*(G;R) \to H_*(G;R)$$

<u>is the identity</u>.

In view of the remarks above, we must show that if h_* is the identity, then h is an inner automorphism. A proof can be constructed by combining special cases of results in [1], [2], and [3]. We shall here present a different, essentially self-contained, argument. This proof is particularly appropriate for a paper in honor of Professor Wilder because it is based on classical results and techniques of algebraic topology - of the sort used so effectively by Professor Wilder and his students.

We first prove the theorem in the special case where G is abelian, that is, a torus. We are given that $h : G \to G$ induces the identity on $H_*(G;R)$. In this case we must prove that h itself is the identity. The Universal Coefficient Theorem produces a commutative diagram

This research was supported in part under NSF Grant No. MCS76-05971.

$$\begin{array}{ccc}
H_1(G;Z)\otimes_Z R & \xrightarrow{h_*^Z \otimes \mathrm{id}} & H_1(G;Z)\otimes_Z R \\
\downarrow \cong & & \downarrow \cong \\
H_1(G;R) & \xrightarrow[h_* = \mathrm{id}]{} & H_1(G;R)
\end{array}$$

where h_*^Z is the endomorphism of integer homology induced by h. Now $H_1(G;Z)$ is free abelian and $h_*^Z \otimes \mathrm{id}$ is the identity on $H_1(G;Z)\otimes_Z R$, so it must be that h_*^Z is itself the identity. The Hurewicz Isomorphism Theorem then implies that $h_\pi : \pi_1(G) \to \pi_1(G)$, the endomorphism of the fundamental group induced by h, is also the identity.

Let $\mathfrak{G}$ be the Lie algebra of G and let $E \subset \mathfrak{G}$ be the kernel of the exponential map. Then E can be identified with $\pi_1(G)$ in such a way that the restriction to E of the differential $dh : \mathfrak{G} \to \mathfrak{G}$ can be identified with h_π. But E spans the vector space $\mathfrak{G}$ so since the linear transformation dh is the identity on E, it is in fact the identity on all of $\mathfrak{G}$. It follows that h is the identity map on G.

Now let G be any compact, connected Lie group and $h : G \to G$ an endomorphism inducing the identity on $H_*(G;R)$. We will next prove that h must be an automorphism. Let $\mathfrak{K}$ be the kernel of the differential $dh : \mathfrak{G} \to \mathfrak{G}$, then $\mathfrak{K}$ is an ideal of $\mathfrak{G}$ and so by [7] a direct summand. Thus inclusion $\mathrm{inc}: \mathfrak{K} \to \mathfrak{G}$ and projection $\mathrm{proj}: \mathfrak{G} \to \mathfrak{K}$ give us a diagram of Lie algebra homomorphisms

$$\begin{array}{ccccc}
 & & \mathfrak{G} & & \\
 & \overset{\mathrm{inc.}}{\nearrow} & & \overset{\mathrm{proj.}}{\searrow} & \\
\mathfrak{K} & & \xrightarrow[\mathrm{id.}]{} & & \mathfrak{K}
\end{array}$$

from which we conclude that

$$(\mathrm{inc})_* : H_*(\mathfrak{K}) \to H_*(\mathfrak{G})$$

is a monomorphism (see [6] for the homology of Lie algebras). Since $h_* : H_*(G;R) \to H_*(G;R)$ is an automorphism, it follows from de Rham's Theorem (see [4]) that $(dh)_* : H_*(\mathfrak{G}) \to H_*(\mathfrak{G})$ is also an automorphism, so $(dh)_*(\mathrm{inc})_* : H_*(\mathfrak{K}) \to H(\mathfrak{G})$ is a monomorphism. But $(dh)(\mathrm{inc}) : \mathfrak{K} \to \mathfrak{G}$ is the constant homomorphism, so $H_j(\mathfrak{K}) = 0$ for $j > 0$ which implies that $\mathfrak{K}$ is of rank zero and therefore trivial. Thus dh is a monomorphism of a finite dimensional vector space, and therefore an automorphism.

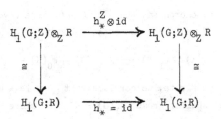

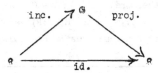

Since dh is an automorphism, then $h : G \to G$ is a covering space with finite fiber K, the kernel of h. From the exact sequence

$$(*) \qquad \pi_1(K) \to \pi_1(G) \xrightarrow{h_\pi} \pi_1(G) \to \pi_0(K) \to \pi_0(G)$$

and the fact that K is finite, we conclude that h_π is a monomorphism. Write the finitely generated abelian group $\pi_1(G) \cong Z^m \oplus \Gamma$ where Z^m is free abelian and Γ is finite. Since h_π is a monomorphism, its restriction to Γ must be an isomorphism from Γ to itself. The argument we used in the abelian case above shows that since h_* is the identity on $H_1(G;R)$, then the restriction of h_π to Z^m is also the identity. We conclude that h_π is an isomorphism. Now $\pi_0(G)$ is trivial, so the exactness of the sequence $(*)$ therefore implies that $\pi_0(K)$ is trivial. Consequently, K is trivial and we have proved that h is an automorphism.

Before we can proceed further with the proof, we must introduce some notation and verify a useful technical result. Let $h : G \to G$ be an endomorphism. Define $h^r : G \to G$ for $r \geq 0$ by letting $h^0(x) = x$ and $h^r(x) = h(h^{r-1}(x))$, for all $x \in G$. For any $k \geq 1$, define a map $\partial^k h : G \to G$ by

$$\partial^k h(x) = x \cdot h(x) \cdot h^2(x) \cdots h^{k-1}(x).$$

LEMMA. *Let $h : G \to G$ be an endomorphism of a compact, connected Lie group of rank ρ such that $h_* : H_*(G;R) \to H_*(G;R)$ is the identity. Then the degree of the map $\partial^k h$ is k^ρ and therefore $\partial^k h$ maps G onto itself.*

Proof. Recall that $\mu : G \times G \to G$ defined by $\mu(x,y) = xy$ induces an algebra structure on $H_*(G;R)$. Furthermore, for $\Delta : G \to G \times G$ the diagonal map, we can choose algebra generators $z_1, \ldots, z_\rho$ for $H_*(G;R)$ so that $\Delta_*(z_j) = 1 \otimes z_j + z_j \otimes 1$ for $j = 1, \ldots, \rho$. An induction argument using the commutative diagram

$$\begin{array}{ccccc} G & & \xrightarrow{\quad \partial^r h \quad} & & G \\ \Delta \downarrow & & & & \uparrow \mu \\ G \times G & \xrightarrow[\mathrm{id} \times \partial^{r-1} h]{} & G \times G & \xrightarrow[\mathrm{id} \times h]{} & G \times G \end{array}$$

proves that $(\partial^k h)_*(z_j) = k z_j$ for $j = 1, \ldots, \rho$. The fact that $z_1 \cdot z_2 \cdots z_\rho$ is the fundamental homology class of G completes the argument.

The next step in the proof of the theorem is to assume that G is a semisimple, compact, connected Lie group and $h : G \to G$ is an endomorphism inducing the identity on $H_*(G;R)$. We showed above that h is in fact an automorphism. Because G is semisimple, there is an integer $m \geq 1$ such that h^m is an inner automorphism, so we can write $h^m(x) = axa^{-1}$ for some $a \in G$ and all x. The

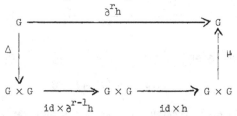

Lemma implies that there exists $b \in G$ such that $\partial^m h(b) = a^{-1}$. Define $h' : G \to G$ by $h'(x) = b \cdot h(x) \cdot b^{-1}$, then h' is homotopic to h by the argument at the beginning of the paper, so h'_* is the identity on $H_*(G;R)$. We observe that, for any $x \in G$,

$$h'^m(x) = (\partial^m h(b)) \cdot axa^{-1} \cdot (\partial^m h(b))^{-1} = x.$$

Choose $t \in G$ with the property that the closure of the subgroup of G generated by t is a maximal torus T of G. Applying the Lemma again, there exists $s \in G$ such that $\partial^m h'(s) = t$. Define $h'' : G \to G$ by $h''(x) = sb \cdot h(x) \cdot b^{-1}s^{-1}$. We claim that $h''(t) = t$; the calculation follows:

$$\begin{aligned} h'(t) &= sb \cdot h(t) \cdot b^{-1}s^{-1} \\ &= sb \cdot h(\partial^m h'(s)) \cdot b^{-1}s^{-1} \\ &= s \cdot h'(\partial^m h'(s)) \cdot s^{-1} \\ &= s \cdot h'(s \cdot h'(s) \ldots h'^{m-2}(s) \cdot h'^{m-1}(s)) \cdot s^{-1} \\ &= s \cdot h'(s) \cdot h'^2(s) \ldots h'^{m-1}(s) \cdot h'^m(s) \cdot s^{-1} \\ &= s \cdot h'(s) \cdot h'^2(s) \ldots h'^{m-1}(s) \\ &= \partial^m h'(s) = t. \end{aligned}$$

The homomorphism h'' therefore has the property $h''(x) = x$ for all x in the subgroup generated by t. Consequently, by continuity, $h''(x) = x$ for all $x \in T$. By [5, Proposition 2.5, p. 334], an automorphism of a semisimple Lie group that is pointwise fixed on a maximal torus is an inner automorphism. Therefore, there exists $c \in G$ such that $h''(x) = cxc^{-1}$ for all $x \in G$. We have shown that

$$sb \cdot h(x) \cdot b^{-1}s^{-1} = c \cdot x \cdot c^{-1}$$

for all x, that is,

$$h(x) = b^{-1}s^{-1}c \cdot x \cdot c^{-1}sb$$

and so h is an inner automorphism.

Now we come to the general case. We have an endomorphism $h : G \to G$ that induces the identity on $H_*(G;R)$. We have shown that h is in fact an automorphism. Write $G = CS$ where C is the component of the center of G containing the unit element and S is connected and semisimple. Since h is an autormophism, the restrictions h_C and h_S are automorphisms of C and S, respectively. The diagram

$$\begin{array}{ccc} H_*(G;R) & \xleftarrow{\cong} & H_*(C;R) \otimes H_*(S;R) \\ {\scriptstyle h_*}\downarrow & & \downarrow{\scriptstyle h_{C*} \otimes h_{S*}} \\ H_*(G;R) & \xleftarrow[\cong]{} & H_*(C;R) \otimes H_*(S;R) \end{array}$$

commutes, so h_{C*} and h_{S*} are the identity on $H_*(C;R)$ and $H_*(S;R)$, respectively. Therefore, by the arguments above, h_C is the identity and h_S is an inner automorphism of S, so h is an inner automorphism.

REFERENCES

1. R. Brown, On the power map in compact groups, Quart. J. Math. **22** (1971), 395-400.
2. ________, On the power map in compact groups, II, Rocky Mt. J. Math. **3** (1973), 9-10.
3. ________, Cohomology of homomorphisms of Lie algebras and Lie groups, Pac. J. Math. **69** (1977), 325-332.
4. C. Chevalley and S. Eilenberg, Cohomology theory of Lie groups and Lie algebras, Trans. Amer. Math. Soc. **63** (1948), 85-124.
5. S. Helgason, Differential Geometry and Symmetric Spaces, Academic Press, 1962.
6. J. Koszul, Homologie et cohomologie des algèbres de Lie, Bull. Soc. Math. France **78** (1950), 65-127.
7. V. Varadarajan, Lie Groups, Lie Algebras, and Their Representations, Prentice-Hall, 1974.

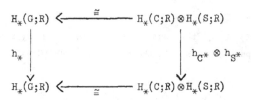

REALIZING HOMOLOGY CLASSES BY PL MANIFOLDS

Gregory W. Brumfiel
Stanford University

Dedicated to R. L. Wilder on the occasion of his 80th birthday.

§1. The Basic Problem

Let $x \in H_k(X,A,\mathbb{Z})$ be a homology class. I will be interested in whether or not there exists a k-dimensional manifold with boundary $M^k, \partial M^k$, and a map of pairs $f: (M,\partial M) \to (X,A)$ with $f_*[M,\partial M] = x$, where $[M,\partial M] \in H_k(M,\partial M,\mathbb{Z})$ is the fundamental class. (All manifolds will be compact and oriented. Boundaries can be empty, e.g., if A is empty.)

This problem has been called the "Steenrod realizability" problem, but I think it is really a much older question in algebraic topology, dating back to the earliest conceptions of homology by Poincaré. In fact, at that time there was some controversy over what Poincaré meant by "cycle" and "homology of a cycle to zero." Certainly, closed manifolds provide the most intuitive pictures of higher dimensional cycles, and manifolds with boundary provide the most intuitive pictures of relative cycles or homologies of a cycle to zero. The actual definition of a homology class as a linear combination of cells with a mixture of algebraic and geometric boundary conditions has proved to be in some ways less subtle but in other ways more subtle than this initial manifold picture. Despite the fundamental nature of this question in algebraic topology, absolutely no progress was made in understanding it until Thom's cobordism paper of 1954, in which he "solved" the smooth manifold version of the problem.

§2. Some Variants of the Basic Problem

In the problem of §1, one could mean by "manifold" either a smooth, combinatorial, or topological manifold, or even a more general class of spaces which carry a fundamental cycle and have some local geometric regularity, such as $\mathbb{Z}$-homology manifolds. It is an easy consequence of the hard work of Kirby and Siebenmann that given a topological manifold $(N^k, \partial N^k)$, there is a PL manifold $(M^k, \partial M^k)$ and a degree one map $f\colon (M^k,\partial M^k) \to (N^k,\partial N^k)$. Thus any homology class $x \in H_k(X,A,\mathbb{Z})$ representable by a TOP manifold is also representable by a PL manifold. Moreover, it is now known (Galewski-Stern, Edwards, Cannon) that given a triangulated $\mathbb{Z}$-homology manifold $(P^k,\partial P^k)$, there is a topological manifold $(N^k,\partial N^k)$ and a degree one map $f\colon (N^k,\partial N^k) \to (P^k,\partial P^k)$. I refer to Frank Raymond's paper in these Proceedings, for a discussion of this point.

Thus, I will concentrate on the differences between the smooth and PL categories. Clearly, if there is a class $x \in H_k(X,A,\mathbb{Z})$ representable by a PL manifold $(N^k,\partial N^k)$ but not by any smooth manifold, then already the fundamental class $[N^k,\partial N^k]$ is not the degree one image of a smooth manifold. Also, the double of a PL manifold $\hat{N} = N \cup_{\partial} (-N)$, is the degree one image of a closed smooth manifold if and only if the pair $(N,\partial N)$ is the degree one image of a smooth pair. This is a simple consequence of transversality; just make any $f\colon M \to \hat{N}$ transverse to the codimension one submanifold $\partial N \subset \hat{N}$. The inverse image separates M into two manifolds with boundary, mapping to the two copies of N in $\hat{N}$. Thus, I will concentrate on closed PL manifolds and ask whether or not they are degree one images of closed smooth manifolds.

This question can be thought of as a sort of resolution of singularities question. It is also a kind of smoothing problem, but note that we are not required to actually smooth our PL manifold, but only its underlying homology class. Thus, for example, a (relatively simple) PL manifold might have the form

$$N^k = W^k \cup_{\partial} L^i \times c\Sigma^{k-i-1}$$

where Σ^{k-i-1} is an exotic sphere, $c\Sigma$ is the cone on Σ and L^i and W^k are smooth manifolds with ∂W^k diffeomorphic to $L^i \times \Sigma^{k-i-1}$. If the cycle $[L^i] \in H_i(N,\mathbb{Z})$ is essential, then it may well carry a non-trivial obstruction to smoothing N. However, since $\Sigma^{k-i-1} = \partial V^{k-i}$ for some smooth manifold V^{k-i}, it is easy to see that N^k is the degree one image of the smooth manifold

$$M^k = W^k \cup_{\partial} L^i \times V^{k-i}.$$

My main result will roughly be that as soon as the smoothing obstructions of N

get sufficiently complicated (in fact, if a certain smoothing obstruction is carried by a homology class not representable by a submanifold of N), then N is not even the degree one image of a smooth manifold.

The key homotopy theoretic fact needed is a formula of Peterson which is ten years old. However, no one seems to have been very interested in this particular consequence of Peterson's formula.

There are some other useful variants of the basic problem. First, for each prime p and $x \in H_k(X,A,\mathbb{Z})$ there is the p-local problem of realizing some multiple μx by a manifold, where $\mu \in \mathbb{Z}$, $(\mu,p) = 1$. Note that if this p-local problem is solved for all p, then by taking a suitable linear combination of p-local solutions, x itself is realized by a manifold.

Secondly, given $n > 0$ and $x \in H_k(X,\mathbb{Z}/n)$, there is the mod-n problem of realizing x as the image of the fundamental class of a $\mathbb{Z}/n$ manifold. A $\mathbb{Z}/n$ manifold is an oriented manifold M with boundary identified with n disjoint copies of an oriented manifold δM, called the Bockstein of M. Gluing these n boundary components together gives a space which carries a fundamental $\mathbb{Z}/n$ homology class. The Bockstein of this class for the coefficient sequence $0 \to \mathbb{Z} \to \mathbb{Z} \to \mathbb{Z}/n \to 0$ is clearly the integral class $[\delta M]$. There is also a relative version of the mod-n problem.

§3. Some results of Thom

I will recall how Thom reformulated the problem of realizing a homology class by a smooth manifold. For simplicity, I consider only the absolute case, $x \in H_k(X,\mathbb{Z})$. Let $V \subset S^{q+k}$ be a (closed) regular neighborhood of X in some sphere, q large, and let $Dx \in H^q(V,\partial V,\mathbb{Z})$ be the Alexander dual of x. Then, as a corollary of transversality, Thom observed that x is realizable by a smooth manifold if and only if there is a map $f\colon V/\partial V \to MSO(q)$ with $f^*U = Dx$, where $U \in H^q(MSO(q),\mathbb{Z})$ is the Thom class.

One of the consequences of the Thom-Milnor-Novikov-Wall study of the algebraic topology of MSO was that for some odd integer μ, a map $f\colon V/\partial V \to MSO(q)$ exists with $f^*U = \mu Dx$. This then solves the 2-local part of the problem of realizing x by a manifold. On the other hand, for odd primes p definite obstructions occur.

For example, if $P^1 \in A_p$ is the first Steenrod operation of degree $2(p-1)$, and β is the Bockstein for the coefficient sequence $0 \to Z \to Z \to Z/p \to 0$, then $\beta P^1 U = 0$ since $H^*(MSO, Z)$ vanishes in dimensions $* \not\equiv 0 \pmod 4$. Thus if $\beta P^1 Dx \neq 0 \in H^*(V/\partial V, Z)$, then no multiple μx is realizable by a manifold if $\mu \not\equiv 0 \pmod p$. In fact, comparing the low dimensional cohomology with that of $K(Z)$ (the Eilenberg-MacLane spectrum) it is not hard to see that for classes $x \in H_k(X, Z)$ with $k \leq 2p+1$, the element $\beta P^1 Dx \in H^*(V/\partial V, Z)$ is the only obstruction to solving the p-local realizability problem for x.

If $k \leq 2p$, this obstruction vanishes. In dimension 2p this requires a little argument. Specifically, if for some x the obstruction didn't vanish, then one would deduce the existence of a non-zero homology operation (the Alexander dual of βP^1), $H_{2p}(X, Z) \to H_1(X, Z)$ with image of order p. Some cohomology class in $H^1(X, Z/p^i)$ would then evaluate non-trivially on the image, producing a map $X \to K(Z/p^i, 1)$, non-zero on the image of the homology operation. But $H_{2p}(K(Z/p^i, 1), Z) = 0$ contradicting naturality of the homology operation. (I give this argument because I will use it again below when I discuss the first obstruction to realizing the fundamental class of a PL manifold by a smooth manifold.) Thus the first dimension in which the p-local realizability problem for homology classes is not always solvable is $2p+1$. If $p = 3$, this gives 7.

Thom's method also applies to the mod-n realizability problem. Let $M(Z/n)$ be the Moore spectrum $\{S^q \cup_n e^{q+1}\}_{q \geq 1}$. Form $MSO \wedge M(Z/n)$ with Z/n Thom class $U_n \in H^*(MSO \wedge M(Z/n))$. Then $x \in H_k(X, Z/n)$ is realizable by a smooth Z/n manifold if and only if there exists $f: V/\partial V \to MSO \wedge M(Z/n)$ with $f^* U_n = Dx$. To identify obstructions, one compares the cohomology of $MSO \wedge M(Z/n)$ with that of the Eilenberg-MacLane spectrum $K(Z/n)$. If $n = p$ is prime, then the first obstruction is $Q_1 Dx \in H^*(V/\partial V, Z/p)$, where $Q_1 \in A_p$ is the Milnor primitive element of degree $2p-1$. (In general, elements Q_i of degree $2p^i - 1$ are defined by calling Q_0 the Bockstein for $0 \to Z/p \to Zp^2 \to Z/p \to 0$, which is the same as the Z/p reduction of β, and setting $Q_{i+1} = [P^{p^i}, Q_i]$.) In the mod-p case, the obstruction can be non-zero in dimension 2p. For example, the non-zero classes in $H_{2p}(K(Z/p,1), Z/p) = Z/p$ cannot be represented by Z/p manifolds. Note that the Bocksteins of these classes are trivially represented by manifolds, namely the Lens spaces.

Once the machinery of PL bundles and transversality was established, it was trivial to restate Thom's results in the PL category. (One can use either PL microbundles or block bundles since the problem is stable.) Thus a class $x \in H_k(X,\mathbb{Z})$ is realizable by a PL manifold if and only if $Dx \in H^*(V/\partial V,\mathbb{Z})$ is the image of the Thom class $U_{PL} \in H^*(MSPL,\mathbb{Z})$ by some map $f\colon V/\partial V \to MSPL$. If, specifically, $x = [N^k]$ is the fundamental class of a PL manifold, then the regular neighborhood V of N can be interpreted as the stable normal bundle of N and $V/\partial V$ is the Thom space $T\nu_N$. The problem of finding a degree one map from a smooth manifold to N thus reduces to the homotopy theory problem of finding $\varphi\colon T\nu_N \to MSO$ such that the diagram below commutes.

3.1

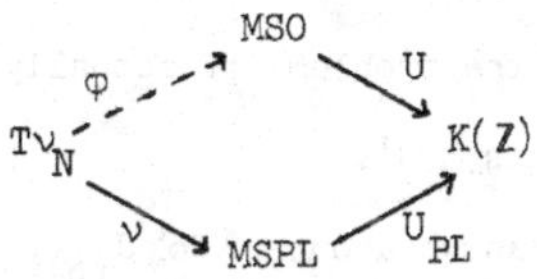

If we fix an odd prime p and look for a degree μ map, $(\mu,p) = 1$, from a smooth manifold to N, then it is necessary and sufficient to solve the same lifting problem with MSO, MSPL, and $K(\mathbb{Z})$ replaced by their localizations $MSO_{(p)}$, $MSPL_{(p)}$, and $K(\mathbb{Z}_{(p)})$.

I mentioned that in §2 that the work of Kirby-Siebenmann implies that every topological manifold is the degree one image of a PL manifold. This follows immediately from the fact that the natural map $MSPL_{(p)} \to MSTOP_{(p)}$ is a homotopy equivalence for all odd primes p.

§4. BSPL and MSPL

In order to study the homotopy theoretic problem stated at the end of §3, it is necessary to study cohomology operations in MSPL. At this date quite a lot is known about $H^*(MSPL)$, due to work of Tsuchiya, Madsen, May, and Mann and Milgram. However, for my purposes all that will be required are some results Sullivan obtained in 1966 and a formula of Peterson dating from 1967.

First, for each odd prime p, Sullivan proved that there is an H-space splitting $BSPL_{(p)} \cong B\,\mathrm{Cok}\,J_{(p)} \times BSO_{(p)}$. Moreover, Sullivan proved that the Thom spectrum factors $MSPL_{(p)} \simeq M\operatorname{Cok}J_{(p)} \wedge MSO_{(p)}$ and that the natural map $BSO_{(p)} \to BSPL_{(p)} = B\operatorname{Cok}J_{(p)} \times BSO_{(p)}$ has the form $0 \times \theta$, where $\theta: BSO_{(p)} \to BSO_{(p)}$ is a certain KO operation. If $\nu: N \to BSPL$ is the stable normal bundle of a PL manifold N, then the obstruction to smoothing N is the obstruction to lifting ν to BSO. The p-local reduction of this obstruction thus breaks into two parts. A certain element of $KO(N) \otimes \mathbb{Z}_{(p)}$ must lie in the image of θ and a certain map $N \to B\operatorname{Cok}J_{(p)}$ must be nullhomotopic. If $\nu_{\operatorname{Cok}J_{(p)}}$ denotes the $B\operatorname{Cok}J_{(p)}$ summand of the stable normal bundle ν, then to find a map $T\nu \to MSO_{(p)}$ solving the p-local smoothing problem for the fundamental homology class of N, it is necessary and sufficient to find a map $T\nu_{\operatorname{Cok}J_{(p)}} \to MSO_{(p)}$ carrying Thom class to Thom class. This discussion provides a sort of qualitative homotopy theoretic distinction between the problems of actually smoothing the manifold N and smoothing the homology class [N].

By now, quite a lot is known about $B\operatorname{Cok}J_{(p)}$. All that I will need is that $\pi_*(B\operatorname{Cok}J_{(p)})$ is isomorphic to the p-torsion of the cokernel of the J-homomorphism (shifted one dimension) $\pi^s_{*-1}/\operatorname{im}(J)$, as module over the stable homotopy groups of spheres π^s_*. In particular, in dimensions up to $4p^2 - 2p - 7$ the only non-zero homotopy groups are $\pi_{2p(p-1)-1} = \mathbb{Z}/p$, with generator β and $\pi_{2p^2-4} = \mathbb{Z}/p$ with generator $\alpha\beta$, where $\alpha \in \pi^s_{2p-3}$ is a generator of the first p-torsion in the stable stem. Since the cohomology operation P^1 is non-zero in $S^q \cup_\alpha e^{q+2p-2}$, it follows that in this range of dimensions $B\operatorname{Cok}J_{(p)}$ is the two stage Postnikov system

$$K(\mathbb{Z}/p, 2p(p-1)-1) \underset{P^1_t}{\times} K(\mathbb{Z}/p, 2p^2-5).$$

The cohomology is easily computed in this range and has $\mathbb{Z}/p$ basis $\{e_1, Q_0e_1, Q_1e_1, Q_0Q_1e_1\}$ where $e_1 \in H^{2p(p-1)-1}(B\operatorname{Cok}J_{(p)}, \mathbb{Z}/p)$ detects β. The reason for the name e_1 is that e_1 is the image of the first exotic class of Gitler-Stasheff $e_1 \in H^{2p(p-1)-1}(BSG, \mathbb{Z}/p)$ under the natural map $B\operatorname{Cok}J_{(p)} \to BSPL_{(p)} \to BSG_{(p)}$. This follows from the fact that the Gitler-Stasheff class also detects β. The Gitler-Stasheff class is defined using twisted secondary operations on Thom classes of spherical fibrations.

The loop space $\operatorname{Cok}J_{(p)} = \Omega B\operatorname{Cok}J_{(p)}$ is a factor of the spaces PL/O =

fibre(BSO $\to$ BSPL) and $SPL_{(p)} = \Omega BSPL_{(p)}$. Later I will construct (tangential) smoothings of manifolds Q by means of mappings $Q \to \mathrm{Cok}\,J_{(p)} \to PL_{(p)} \to PL/O$. If Q has sufficiently small dimension, then $\mathrm{Cok}\,J_{(p)}$ looks like a two stage Postnikov system to Q and there is an exact sequence

$$H^{2p(p-1)-3}(Q,Z/p) \xrightarrow{P^1}$$

$$H^{2p^2-6}(Q,Z/p) \to [Q, \mathrm{Cok}\,J_{(p)}] \xrightarrow{\sigma e_1} H^{2p(p-1)-2}(Q,Z/p) \xrightarrow{P^1} H^{2p^2-5}(Q,Z/p),$$

where $\sigma e_1 \in H^{2p(p-1)-2}(\mathrm{Cok}\,J_{(p)}, Z/p)$ is the desuspension of $e_1 \in H^{2p(p-1)-1}(B\,\mathrm{Cok}\,J_{(p)}, Z/p)$.

§5. The Formula of Peterson

The crucial formula needed to find the first examples of PL manifolds N not the degree one image of smooth manifolds is the following

5.1 $$Q_2U_{PL} = \Phi(Q_0P^1Q_0e_1) \in H^{2p^2-1}(MSPL,Z/p),$$

where $Q_2 \in A_p$ is the Milnor primitive operation of degree $2p^2 - 1$ defined in §3 and $\Phi\colon H^*(BSPL,Z/p) \xrightarrow{\approx} \tilde{H}^*(MSPL,Z/p)$ is the Thom isomorphism. (Strictly speaking, 5.1 may be correct only up to unit in Z/p.) It is an immediate corollary that if N is a PL manifold such that $Q_0P^1Q_0e_1(N) \neq 0 \in H^{2p^2-1}(N,Z/p)$, where $e_1(N) = \nu^*(e_1)$, $\nu\colon N \to BSPL$ the normal bundle, then N is not the degree one image of a smooth manifold. Namely, the lifting problem of diagram 3.1 cannot be solved since $Q_2U = 0 \in H^{2p^2-1}(MSO,Z/p)$.

The existence of such PL manifolds N is an easy matter. By thickening a skeleton of BSPL in Euclidean space and then taking the total space of a universal PL bundle over this thickening, one obtains a PL manifold N (with boundary) whose normal bundle is a universal PL bundle for low dimensional spaces. Since $Q_0P^1Q_0e_1 \neq 0 \in H^*(BSPL,Z/p)$, we will have $Q_0P^1Q_0e_1(N) \neq 0$ also. However, the lowest dimensional example obtainable by this method would probably have dimension around 60.

The actual obstructions to solving the lifting problem $3.1_{(p)}$ are p-torsion elements in $H^{4*+1}(N,Z)$. In fact, the obstructions for the p-local problem can be specifically identified as the k-invariants of the Brown-Peterson spectrum $BP_{(p)}$, applied to the Thom class $U \in H^*(T\nu_N, Z)$. This is so because $MSO_{(p)}$ is a sum of suspensions of $BP_{(p)}$. In §4, I described $H^*(B\,Cok\,J_{(p)}, Z/p)$ in a range of dimensions. This also computes $H^*(BSPL_{(p)}, Z/p)$ and $H^*(BSPL_{(p)}, Z)$ in this range using $BSPL_{(p)} \simeq B\,Cok\,J_{(p)} \times BSO_{(p)}$. In particular, the first non-zero element of $H^{4*+1}(BSPL_{(p)}, Z)$ is $\beta P^1 \beta e_1 \in H^{2p^2-1}(BSPL_{(p)}, Z)$. Recall Q_0 is β reduced mod p. Thus, from Peterson's formula 5.1, $\beta P^1 \beta e_1(N) \in H^{2p^2-1}(N,Z)$ really is the first obstruction to lifting in $3.1_{(p)}$, and, in fact, is the only obstruction if $\dim(N) \le 2p^2 + 1$ or even a little higher.

I claim that if $\dim(N) \le 2p^2$, then $\beta P^1 \beta e_1(N) = 0$. If $\dim(N) < 2p^2$, this is trivial. If $\dim(N) = 2p^2$, repeat Thom's argument given in §3 proving some multiple μx of a 2p dimensional homology class x, with $(\mu,p) = 1$, is always realizable by a manifold. The main point is that $H_{2p^2}(K(Z/p^i,1),Z) = 0$, so no homology operation with image of order p can be non-zero from $H_{2p^2}(X,Z)$ to $H_1(X,Z)$ for any space X. Thus the smallest dimension in which there can exist a PL manifold N, not the degree μ image of a smooth manifold for any μ with $(\mu,p) = 1$, is $2p^2 + 1$. For $p = 3$, this gives 19.

Note that the formula for the obstruction $\beta P^1 \beta e_1(N) = \beta P^1(\beta e_1(N))$ is exactly the first obstruction to finding a map $f: N^k \to MSO(2p(p-1))$ with $f^*U = \beta e_1(N)$. This is precisely the obstruction to realizing the Poincaré dual $D\beta e_1(N) \in H_{k-2p(p-1)}(N,Z)$ by an embedded submanifold in N. If $k = 2p^2 + 1$, then $2p + 1 = k - 2p(p-1)$ and this is precisely the first dimension in which this obstruction can be non-zero. Note that $e_1(N)$ and $\beta e_1(N)$ are exotic characteristic classes of the PL normal bundle of N and hence are certainly smoothing obstructions. In fact, in the $B\,Cok\,J_{(p)}$ part of the total smoothing obstruction discussed in §4, $e_1(N) \in H^{2p(p-1)-1}(N,Z/p)$ is the first obstruction. Recall I observed in §4 that this $B\,Cok\,J_{(p)}$ part of the normal bundle is all that is relevant as far as smoothing the homology class [N] is concerned. Thus, what has been argued above is, roughly, that as soon as the homology class $De_1(N) \in H_{k-2p(p-1)+1}(N,Z/p)$ is sufficiently complicated, in the sense that its Bockstein $D\beta e_1(N)$ is not represented by any submanifold of N, then N is not a degree μ image of a smooth manifold, for any μ with $(\mu,p) = 1$.

There is perhaps a little ambiguity here about what one means by a "submanifold of N". If the submanifold is produced by $N \to MSO(2p(p-1))$, then one obtains a PL submanifold with smooth normal bundle in N. If the submanifold is produced instead by $T\nu_N^q \to MSO(q + 2p(p-1))$, then one obtains a smooth submanifold of ν_N^q. If the dimension of N is sufficiently high, certainly $2p^2 + 1$ suffices, this smooth manifold can then be embedded in N, but perhaps with non-smooth normal bundle in N. These two approaches to realizing $D\beta e_1(N)$ by a submanifold of N are entirely equivalent.

§6. An Example in Dimension $2p^2 + 1$

One method of constructing PL manifolds is to glue together two smooth manifolds with boundary by means of a PL isomorphism between their boundaries. Slightly more generally, suppose Q is any closed smooth manifold and suppose $f\colon Q' \to Q$ is a smoothing of Q. That is, Q' is also a smooth manifold and f is a PL isomorphism. Then the Pontrjagin numbers and Stiefel-Whitney numbers of Q and Q' coincide, hence Q and Q' are smoothly cobordant. If V is a smooth manifold with $\partial V = Q \cup (-Q')$, then one obtains a closed PL manifold $N = V/\sim$, where $\sim$ means identify Q and Q' using $f\colon Q' \to Q$. In case Q and Q' are already boundaries, say $Q = \partial W$, $Q' = \partial W'$, we can form $N = W' \cup_f W$.

I want to construct such N of dimension $2p^2 + 1$ with $\beta P^1 \beta e_1(N) \neq 0$. Thus Q will have dimension $2p^2$. Let $W_0 \subset W$ denote W minus an open collar neighborhood of $\partial W = Q$ and similarly let $W_0' \subset W'$ denote W' minus a collar of Q'. Thus we can write $W' = W_0' \underset{Q \times \{0\}}{\cup} Q' \times [0,1/2]$ and $W = W_0 \underset{Q \times \{1\}}{\cup} Q \times [1/2,1]$. The normal bundle $\nu\colon N \to BSPL$ fits into a diagram

$$\begin{array}{ccc} W_0' \amalg W_0 & \longrightarrow & BSO \\ \downarrow & & \downarrow \\ N & \xrightarrow{\nu} & BSPL \\ \rho\downarrow & & \downarrow\pi \\ (N, W_0' \amalg W_0) & \xrightarrow{\bar{\nu}} & (BSPL, BSO). \end{array}$$

Since e_1 is naturally a relative class, $e_1 \in H^{2p(p-1)-1}(BSPL, BSO, \mathbb{Z}/p)$, we have $e_1(N) = \nu^*\pi^*(e_1) = \rho^*\bar{\nu}^*(e_1)$ and also, of course, $\beta P^1\beta e_1(N) = \rho^*\bar{\nu}^*(\beta P^1\beta e_1)$.

Now, the concordance class of the smoothing $f\colon Q' \to Q$ is classified by a homotopy class of maps $\varphi\colon Q \to PL/O$. The space PL/O, which is the fibre of $BSO \to BSPL$, classifies stable concordance classes of vector bundles together with PL trivializations. An equivalent way of viewing a homotopy class $\varphi\colon Q \to PL/O$ is as a certain equivalence class of PL bundles γ, over $Q \times I$, together with a PL bundle identification of $\gamma|_{Q \times \{0\}}$ with a vector bundle and a PL bundle identification of $\gamma|_{Q \times \{1\}}$ with the trivial bundle. This is just the universal map of pairs $\alpha\colon (PL/O \times I, PL/O \times \dot{I}) \to (BSPL, BSO)$, which is the inclusion $PL/O \to BSO$ on $PL/O \times \{0\}$ together with a null-homotopy of PL/O in $BSPL$.

Consider the excision map $h\colon (Q \times I, Q \times \dot{I}) \to (N, W_0' \amalg W_0)$ defined by the diagram below.

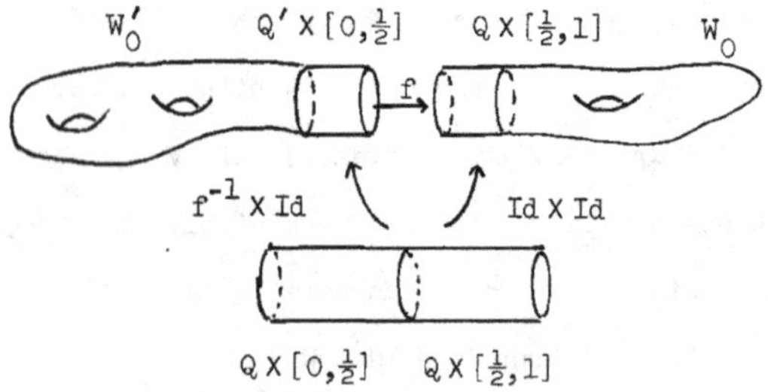

$$\begin{array}{c} N \\ \uparrow h \\ Q \times I \end{array}$$

If we pull the PL tangent bundle of N back along h and subtract (stably) the tangent bundle of $Q \times I$, we get a PL bundle over $Q \times I$, trivial on $Q \times \{1\}$ and identified with a vector bundle (the difference of the tangent bundles of Q' and Q) on $Q \times \{0\}$. But this is precisely the mechanism by which smoothing theory establishes a correspondence between concordance classes of smoothings of Q and homotopy classes of maps $[Q, PL/O]$. In other words, the excision h fits into a commutative diagram.

$$\begin{array}{ccc} N & \xrightarrow{\nu} & BSPL \\ \rho\downarrow & & \downarrow\pi \\ (N, W_0' \amalg W_0) & \xrightarrow{\bar{\nu}} & (BSPL, BSO) \\ h\uparrow & & \uparrow\alpha \\ (Q \times I, Q \times \dot{I}) & \xrightarrow{\varphi \times Id} & (PL/O \times I, PL/O \times \dot{I}). \end{array}$$

This tells us that $\beta P^1\beta e_1(N) = \rho^*(h^*)^{-1}(\varphi \times \mathrm{Id})^*\alpha^*(\beta P^1\beta e_1)$.

A simple analysis of the exact cohomology sequence of the pair $(N, W_0' \amalg W_0)$ and the excision isomorphism h^* induced by $h: (Q \times I, Q \times \dot{I}) \to (N, W_0' \amalg W_0)$ then shows that $\beta P^1\beta e_1(N) \neq 0$ if and only if $\varphi^*(\beta P^1\beta(\sigma e_1)) \notin$ image$(H^*(W' \amalg W, Z) \xrightarrow{(f^{-1})^*(i')^*+i^*} H^*(Q,Z))$ where $\sigma e_1 \in H^{2p(p-1)-2}(PL/O, Z/p)$ is the desuspension of $\alpha^* e_1 \in H^{2p(p-1)-1}(PL/O \times I, PL/O \times \dot{I})$ and $i: Q \to W$ and $i': Q' \to W'$ are the inclusions of boundaries. We will make our choice of Q^{2p^2} and W^{2p^2+1} so that, in fact, $H^{2p^2-1}(W,Z) = 0$. Then we need to choose $f: Q' \to Q$ and $Q' = \partial W'$ so that $f^*\varphi^*(\beta P^1\beta(\sigma e_1)) \notin$ image$(H^*(W',Z) \to H^*(Q',Z))$. The Poincaré dual statement is that the integral homology class of order p, $Df^*\varphi^*\beta P^1\beta(\sigma e_1) \in H_2(Q',Z)$ does not bound in W'. It is thus necessary and sufficient to find a cohomology class $z \in H^2(Q',Z/p^i)$, some $i \geq 1$, which evaluates non-trivially on $Df^*\varphi^*\beta P^1\beta(\sigma e_1)$ and such that the map $z: Q' \to K(Zp^i,2)$ is smoothly bordant to zero, through some $W' \to K(Z/p^i,2)$. It is a somewhat disturbing but unaviodable complication that although Q and Q' are "identified" by f, they are different as smooth manifolds. Hence the fact that $z: Q \to K(Z/p^i,2)$ bounds $W \to K(Z/p^i,2)$ is of little help in deciding if $z: Q' \to K(Z/p^i,2)$ is zero as smooth bordism element. What is obvious is that (Q',z) is zero as PL bordism element, in other words, $(Q',z) \in \mathrm{kernel}(\tilde{\Omega}_{2p^2}(K(Z/p^i,2)) \to \tilde{\Omega}^{PL}_{2p^2}(K(Z/p^i,2))$.

To find a suitable Q^{2p^2}, then, our first requirement is that there should exist $\varphi: Q \to PL/O$ with $\varphi^*(\beta P^1\beta(\sigma e_1)) \neq 0$. As observed at the end of §4, the space $\mathrm{Cok}\,J_{(p)}$ is a factor of SPL and PL/O, and we have maps $\varphi: Q \to \mathrm{Cok}\,J_{(p)}$ with $\varphi^*(\sigma e_1) = u$ for any class $u \in H^{2p(p-1)-2}(Q,Z/p)$ with $P^1 u = 0$. As an example with $\beta P^1\beta u \neq 0$ we can take $Q = S^{2p(p-1)-4} \times L^3 \times L^{2p+1}$ and $u = \sigma \otimes x \otimes x$, where L^{2j+1} is a Lens space $\frac{S^{2j+1}}{Z/p}$, and $\sigma \in H^{2p(p-1)-4}(S^{2p(p-1)-4},Z)$ and $x \in H^1(L^{2j+1},Z/p)$ are generators. In this case, $Q_0P^1Q_0u = \sigma \otimes y \otimes y^p$, where $y = Q_0x \in H^2(L^{2j+1},Z/p)$. Also, this choice of Q satisfies our second requirement that Q bound a W of low dimensional homotopy type. Specifically, we can take $W = D^{2p(p-1)-3} \times L^3 \times L^{2p+1}$. The class $\beta P^1\beta u$ is Poincaré dual to the torus $pt \times S^1 \times S^1 \subset S \times L^3 \times L^{2p+1}$ and the cohomology class $1 \otimes x \otimes x \in H^2(S \times L^3 \times L^{2p+1},Z/p)$ evaluates non-trivially.

Thus, choose some $\varphi: Q \to \mathrm{Cok}\,J_{(p)} \to PL$ with $\varphi^*(\sigma e_1) = \sigma \otimes x \otimes x$, giving a smoothing $f: Q' \to Q$. Geometrically, Q' has the same smooth structure as Q on the $2p(p-1)-3$ skeleton. The $2p(p-1)-2$ skeleton contains the smooth submanifold $(S^{2p(p-1)-4} \times S^1 \times S^1) \# \Sigma^{2p(p-1)-2}$, where the exotic sphere $\Sigma^{2p(p-1)-2}$ corresponds to β in the isomorphism $\Gamma_{2p(p-1)-2} \simeq \pi^S_{2p(p-1)-2}/\mathrm{im}(J)$. The smoothing is then extended up the skeletons of $S^{2p(p-1)-4} \times L^3 \times L^{2p+1}$. But we now come to the delicate question of whether $(1 \otimes x \otimes x) \circ f: Q' \to Q \to K(Z/p,2)$ bounds some $W' \to K(Z/p,2)$. I see no direct geometric construction which proves this. However, the desired result does follow from some rather involved computations of $\tilde{\Omega}_*(K(Z/p,2))$ and the homomorphism $\tilde{\Omega}_*(K(Z/p,2)) \to \tilde{\Omega}^{PL}_*(K(Z/p,2))$. This argument will be deferred until the next section. Assuming this result, we have succeeded in constructing $N = W' \cup_f W$, a PL manifold of dimension $2p^2 + 1$, which is not the degree μ image of a smooth manifold if $\mu \not\equiv 0$ (mod. p).

If one just wants a homology class realizable by a PL manifold but not by a smooth manifold, presumably less work is required. In fact, a suitable class will exist in $H_{2p^2+1}(K(Z/p,2),Z)$. It should suffice to find a class on which $Q_2\iota$ evaluates non-trivially, but on which cohomology elements in the image of Q_0 and Q_1 evaluate trivially. Our manifold N above will represent such a class if maps $W \to K(Z/p,2)$ and $W' \to K(Z/p,2)$ are fit together along $1 \otimes x \otimes x$: $Q \to K(Z/p,2)$.

Perhaps even a simpler class representable by a PL manifold but not by a smooth manifold is the torsion product $[L^{2p^2-1}] * [S^1] \in H_{2p^2+1}(L \times L, Z)$, where $L = K(Z/p,1)$ is the infinite Lens space. This fits in nicely with the fact that the simplest example of a homology class not represented by any kind of manifold is the torsion product $[L^{2p-1}] * [S^1] \in H_{2p+1}(L \times L, Z)$.

Perhaps surprisingly, if one considers Z/p homology classes and asks for the lowest dimensional class representable by a PL Z/p manifold but not by a smooth Z/p manifold, the answer is still $2p^2 + 1$, not $2p^2$. The reason is that although the homology operation dual to Q_2 can be non-zero $H_{2p^2}(X,Z/p) \to H_1(X,Z/p)$, Peterson's formula 5.1 ties together Q_2 and Q_0Q_1, and the dual of the composition Q_0Q_1 is always zero, $H_{2p+2}(X,Z/p) \to H_1(X,Z/p)$, because of properties of $K(Z/p,1)$.

§7. A Technical Argument

I will give the steps of a proof that $(1 \otimes x \otimes x) \circ f\colon Q' \to Q \to K(Z/p,2)$ is smoothly bordant to zero, where $Q = S^{2p(p-1)-4} \times L^3 \times L^{2p+1}$ and $f\colon Q' \to Q$ is a smoothing of Q induced by $\varphi\colon Q \to \operatorname{Cok} J_{(p)} \to PL$ with $\varphi^*(\sigma e_1) = \sigma \otimes x \otimes x$, $x \in H^1(L, Z/p)$ a generator. I benefited from conversations with Jim Milgram while working out the argument. Unfortunately the argument is extremely technical. There are hopefully better ways of accomplishing the final goal, which was, don't forget, to give some "construction" of a PL manifold not the degree one image of a smooth manifold.

First, for any space, X, $\tilde{\Omega}_*(X) = \pi_*(X \wedge MSO)$. If X is a p-torsion space, like $K(Z/p,2)$, then $\pi_*(X \wedge MSO) = \pi_*(X \wedge MSO_{(p)}) = \oplus_\alpha \pi_*(X \wedge BP_{(p)} \wedge S^{n_\alpha})$, since $MSO_{(p)} \simeq \bigcup_\alpha BP_{(p)} \wedge S^{n_\alpha}$. Here $BP_{(p)}$ is the Brown-Peterson spectrum with $H^*(BP_{(p)}, Z/p) = A_p/(Q_0) = A_p/A_p\{Q_0, Q_1, Q_2, \ldots\}$, the algebra generated by the Steenrod p^{th} power operations. From now on we suppress all symbols localizing spaces at p, as this will be understood. BP is a ring spectrum and we have $\pi_*(BP) = Z_{(p)}[v_1, v_2, \ldots]$ where the degree of v_i is $2(p^i - 1)$.

Secondly, a convenient method (sometimes) for computing $\pi_*[BP \wedge X]$ is the Adams spectral sequence. A "change of rings" type result is that

$$\operatorname{Ext}_{A_p}(A_p/(Q_0) \otimes H^*(X), Z/p) = \operatorname{Ext}_{E(Q_0,Q_1,\ldots)}(H^*(X), Z/p)$$

where $E(Q_0, Q_1, \ldots)$ is the exterior algebra on the Q_i.

If $X = K(Z/p,2)$, then $H^*(X) = P(\iota, Q_0Q_1\iota, Q_0Q_2\iota, \ldots) \otimes E(Q_0\iota, Q_1\iota, Q_2\iota, \ldots)$, where P is a polynomial algebra, E is an exterior algebra, and $\iota \in H^2(K(Z/p,2))$ is the fundamental class. The action of the Q_i can be determined, but in dimensions up to $2p^2 + 1$ this is especially simple since the Q_i, $i \geq 2$, can be ignored for dimensional reasons. The result is that in dimensions up to $2p^2 + 1$, $H^*(K(Z/p,2))$ is a sum of (a) free modules over $E(Q_0, Q_1, \ldots)$ on even dimensional generators, (b) trivial modules Z/p on generators $\iota^p, \iota^{2p}, \ldots, \iota^{p^2}$, and (c) modules with two generators, $\iota^{jp-1} \cdot Q_0\iota$ and $\iota^{jp-1} \cdot Q_1\iota$, related by $Q_0(\iota^{jp-1} \cdot Q_0\iota) = Q_1(\iota^{jp-1} \cdot Q_1\iota) = 0$ and $Q_1(\iota^{jp-1} \cdot Q_0\iota) = \pm Q_0(\iota^{jp-1} \cdot Q_1\iota)$. The E_2 term of the

Adams spectral sequence for $\pi_*(BP \wedge K(Z/p,2))$, which is $\mathrm{Ext}_{E(Q_0,Q_1,\dots)}(H^*(K(Z/p,2)),Z/p)$, is then easily computed.

The modules of type (b) and (c) pair up and give p copies of the following diagram, or part of it, beginning at the left, until dimensions exceed $2p^2 + 1$.

(*)

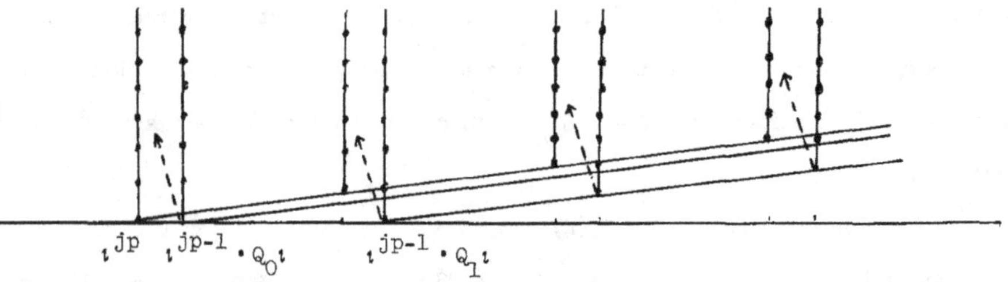

The dots indicate Z/p, vertical lines denote Q_0-towers, and angled lines denote Q_1-towers (which are only indicated at the bottom but which do extend up to Q_0 towers). The dashed arrows indicate differentials. They are somewhat ambiguously drawn since if $1 \le j \le p - 1$, I mean $d_2 \neq 0$ and if $j = p$, I mean $d_2 = 0$ but $d_3 \neq 0$. At the E_∞-level, Q_0-towers correspond to multiplication by 3 and Q_1-towers correspond to multiplication by $v_1 \in \pi_{2(p-1)}(BP)$. These operations commute with differentials. Thus the only possible pattern of differentials is forced by what happens to $\iota^{jp-1} \cdot Q_0\iota$ and this is determined by the higher order Bockstein relation between ι^{jp} and $\iota^{jp-1} \cdot Q_0\iota$. Specifically, if $1 \le j \le p - 1$, then $\beta_2(\iota^{jp}) = \iota^{jp-1}Q_0\iota$ (up to unit in Z/p), and $\beta_3(\iota^{p^2}) = \iota^{p^2-1} \cdot Q_0\iota$. The resulting elements of order p^2 and p^3 in $\pi_*(K(Z/p,2) \wedge BP)$ are detected by cohomology classes with Z/p^2 and Z/p^3 coefficients, namely the Pontrjagin p^{th} powers of ι^j, $1 \le j \le p - 1$, and the Pontrjagin p^2-power of ι. Geometrically, back in $\Omega_*(K(Zp,2))$, these elements are represented by the maps z^{jp}: $CP(jp) \to K(Z/p,2)$.

The free module generators of type (a) in $H^*(K(Z/p,2))$ also give rise to elements in $\pi_*(K(Z/p,2) \wedge BP)$ detected by cohomology classes (with Z/p coefficients). Inspection of diagram (*) shows that the kernel of the Hurewicz homomorphism $\pi_*(BP \wedge K(Z/p,2)) \to H_*(K(Z/p,2),Z)$, in dimensions up to $2p^2$, is a direct sum of groups Z/p on generators connected to the ι^{jp} by multiplication by powers of the generator $v_1 \in \pi_{2(p-1)}(BP)$. This is as much information as we need about $\pi_*(BP \wedge K(\mathbb{Z}/p,2))$.

Next, we must study the homotopy theoretic interpretation of the effect on a bordism element $g: Q \to X$ of a change of smooth structure on Q, $g \circ f: Q' \to Q \to X$. We will assume the smoothing $f: Q' \to Q$ corresponds to a map $\varphi: Q \to PL$. The case $\varphi: Q \to PL/O$ is slightly more complicated, but needn't concern us here.

Regard PL as a big group of base point preserving PL isomorphisms of a sphere S, with action map $ev: S \wedge PL_+ \to S$. Given $g: Q \to X$ and $\varphi: Q \to PL$ we get a bordism element $\varphi \times g: Q \to PL \times X$, which is a homotopy element of $S \wedge PL_+ \wedge MSO \wedge X_+$. The effect on bordism of change of smoothing is just the map

7.1 $$\pi_*(S \wedge PL_+ \wedge MSO \wedge X_+) \xrightarrow{ev \wedge Id \wedge Id} \pi_*(S \wedge MSO \wedge X_+).$$

Geometrically, the suspension $Id \times g: S \times Q \to S \times X$ gets replaced by $ev(\varphi) \times g: S \times Q \to S \times X$, where $\varphi: Q \to PL$. Now when $ev(\varphi) \times g$ is made smoothly transverse to $pt \times X \subset S \times X$ (which is the geometric construction of the bordism suspension isomorphism) one gets $g \circ f: Q' \to X$. In particular, the map 7.1 breaks up into a direct sum of maps, corresponding to $MSO = \bigcup_\alpha BP \wedge S^{n_\alpha}$.

We can replace PL in this discussion by its subspace (subgroup) Cok J, and we can specialize to $X = K(\mathbb{Z}/p,2)$. Consider the bordism element $\gamma = \varphi \times (1 \otimes x \otimes x): Q \to \text{Cok J} \times K(\mathbb{Z}/p,2)$ where $Q = S^{2p(p-1)-4} \times L^3 \times L^{2p+1}$ and $\varphi^*(\sigma e_1) = \sigma \otimes x \otimes x$. Now, $S \wedge \text{Cok J}_+ = S \cup (S \wedge \text{Cok J})$ and the map in 7.1 is clearly the identity on the factor $\pi_*(S \wedge MSO \wedge X_+)$. Thus we are only interested in the image of the bordism class of (Q,γ) in the factor $\pi_*(S \wedge \text{Cok J} \wedge MSO \wedge K(\mathbb{Z}/p,2))$, which we will denote $(Q,\bar{\gamma})$.

In dimensions up to $2p^2 + 1$, we have that $H^*(\text{Cok J})$ is free over $E(Q_0,Q_1,\dots)$ on $\sigma e_1 \in H^{2p(p-1)-2}(\text{Cok J})$. Thus $H^*(\text{Cok J} \wedge K(\mathbb{Z}/p,2))$ is also free over $E(Q_0,Q_1,\dots)$ in this range, on generators $\{\sigma e_1 \otimes u_i\}$, where $\{u_i\}$ is a $\mathbb{Z}/p$ base of $H^*(K(\mathbb{Z}/p,2))$. Hence all elements of $\pi_*(S \wedge \text{Cok J} \wedge BP \wedge K(\mathbb{Z}/p,2))$ are of order p and are detected by cohomology classes $\{\sigma e_i \otimes 1 \otimes u_i\}$, $1 \in \tilde{H}^0(BP)$ the unit (as ring spectrum, not space). It follows that all bordism elements $h: M \to \text{Cok J} \wedge K(\mathbb{Z}/p,2)$ are detected by $\mathbb{Z}/p$ cohomology characteristic numbers $\langle y_\alpha(M) \cdot h^*(\sigma e_i \otimes u_i),(M)\rangle$, where $\Phi y_\alpha \in H^*(MSO,\mathbb{Z}_{(p)})$ are the images of the units in $BP \wedge S^{n_\alpha}$ under $MSO = \bigcup_\alpha BP \wedge S^{n_\alpha}$.

The Lens space L^3 is parallelizable and the tangent bundle of L^{2p+1} is computable from the fibration $S^1 \to L^{2p+1} \to CP(p)$, which is the complex line bundle with Chern class $pz \in H^2(CP(p),Z)$. It turns out simply that $p_1(L^{2p+1}) \neq 0$ and $p_i(L^{2p+1}) = 0$, $i > 1$, but, regardless, it is easy to show that all cohomology characteristic numbers of (Q,γ) of the form $\langle P_I(Q) \cdot \gamma^*(v),[Q]\rangle$ vanish, if P_I is a non-trivial monomial in the Pontrjagin classes and $v \in H^*(\mathrm{Cok}\,J \times K(Z/p,2),Z/p)$. The conclusion is that the bordism class $(Q,\bar{\gamma}) \in \pi_*(S \wedge \mathrm{Cok}\,J \wedge MSO \wedge K(Z/p,2))$ lives in the bottom summand $\pi_*(S \wedge \mathrm{Cok}\,J \wedge BP \wedge K(Z/p,2))$ and hence its image by 7.1 in $\pi_*(S \wedge MSO \wedge K(Z/p,2))$ also lives in the bottom summand $\pi_*(S \wedge BP \wedge K(Z/p,2))$.

Since $H^{2p+2}(K(Z/p,2))$ is spanned by ι^{p+1} and $Q_0Q_1\iota$, the class $(Q,\bar{\gamma})$ is determined by the formulas $\langle\gamma^*(\sigma e_1 \otimes Q_0Q_1\iota),[Q]\rangle = 1$ and $\langle\gamma^*(\sigma e_1 \otimes \iota^{p+1}),[q]\rangle = 0$. In particular, $(Q,\bar{\gamma}) \neq 0$, so we must consider more carefully the possible images in $\pi_*(S \wedge BP \wedge K(Z/p,2))$. Certainly the image is in the kernel of the Hurewicz homomorphism. But in our computation of $\pi_*(BP \wedge K(Z/p,2))$ above we specifically found some non-zero elements in this Hurewicz kernel. Even so, we would be finished if we knew multiplication by $v_1 \in \pi_{2(p-1)}(BP)$ was injective on the Hurewicz kernel in $\pi_{2p^2}(BP \wedge K(Z/p,2))$, since multiplication by v_1 does kill $(Q,\bar{\gamma})$. Unfortunately, this would involve computation up to $\pi_{2p^2+2(p-1)}(BP \wedge K(Z/p,2))$ and the $E(Q_0,Q_1,\ldots)$-module structure of $H^*(K(Z/p,2))$ just gets sufficiently complicated in this dimension to make the Adams spectral sequence differentials ambiguous, at least to me. Some other eager beaver might push through this approach, however.

The one thing we do know about the bordism element $(1 \otimes x \otimes x) \circ f$: $Q' \to Q \to K(Z/p,2)$ is that it is zero as PL bordism element. This implies that our image of $(Q,\bar{\gamma})$ in $\pi_*(S \wedge MSO \wedge K(Z/p,2))$ belongs to kernel $(\pi_*(MSO \wedge K(Z/p,2)) \to \pi_*(MSPL \wedge K(Z/p,2)))$. Now, we have Sullivan's result $MSPL = M\,\mathrm{Cok}\,J \wedge MSO$. If $j\colon MSO \to MSPL$ is the natural map, then j is rather complicated. However, if $i_0\colon BP \to MSO$ and $\rho_0\colon MSO \to BP$ are inclusion and projection respectively of the bottom summand BP of MSO, then the composition

$k = (\mathrm{Id} \wedge \rho_0) j\, i_0 \colon BP \to MSO \to MSPL = M\,\mathrm{Cok}\,J \wedge MSO \to M\,\mathrm{Cok}\,J \wedge BP$ can be identified after automorphism of BP with the identity $BP \to S^0 \wedge BP$, where $S^0 \to M\,\mathrm{Cok}\,J$ is the Thom cell. This fact lies deeper in Sullivan's work and is too complicated to even outline here. The significance, however, is we know we are dealing with an element of the bottom summand $\pi_*(BP \wedge K(\mathbb{Z}/p,2))$ of $\pi_*(MSO) \wedge K(\mathbb{Z}/p,2))$, so we will be finished if we show that the map $k_* \colon \pi_*(BP \wedge K(\mathbb{Z}/p,2)) \to \pi_*(M\,\mathrm{Cok}\,J \wedge BP \wedge K(\mathbb{Z}/p,2))$ is injective in dimensions $\leq 2p^2$. This is exactly within reach because of the simple form of $k \colon BP \to M\,\mathrm{Cok}\,J \wedge BP$ and some further computations, basically due to Peterson again, in the same paper in which he proved 5.1.

Specifically, Peterson computed $\pi_*(M\,\mathrm{Cok}\,J \wedge BP)$ for a range of dimensions considerably beyond $2p^2$, by studying $H^*(M\,\mathrm{Cok}\,J)$ over $E(Q_0, Q_1, \ldots)$. In dimensions up to $2p^2 + 1$, $H^*(M\,\mathrm{Cok}\,J)$ has generators U, e_1U, $Q_0(e_1U)$, $Q_1(e_1U)$ and $Q_2U = Q_0Q_1(e_1U)$, with $Q_0U = Q_1U = 0$. It is then an easy Adams spectral sequence computation that the map $S \wedge BP \to M\,\mathrm{Cok}\,J \wedge BP$ induces a monomorphism in π_* (through $2p^2 + 1$, at least) with cokernel $\mathbb{Z}/p$ in dimension $2p(p-1) - 1$ detected by $e_1U \otimes 1 \in H^{2p(p-1)-1}(M\,\mathrm{Cok}\,J \wedge BP)$.

Recall $\pi_*(BP) = \mathbb{Z}_{(p)}[v_1, v_2, \ldots]$ where degree $(v_i) = 2(p^i - 1)$. In dimensions up to $2p^2$ we have generators v_1 and v_2 and $v_1^2, v_1^3, \ldots, v_1^{p+1}$. Comparing BP with $M\,\mathrm{Cok}\,J \wedge BP$ as above, and using 5.1, we see that in dimensions up to $2p^2$, $M\,\mathrm{Cok}\,J \wedge BP$ has the same k-invariants as BP, except that in dimension $2p(p-1) - 1$ a $K(\mathbb{Z}/p,\, 2p(p-1) - 1)$ appears with trivial k-invariant, giving rise to the cohomology element $e_1U \otimes 1$, and in dimension $2p^2 - 2$ the k-invariant corresponding to the homotopy generator $v_2 \in \pi_{2p^2-2}(BP)$ is the sum of the original BP k-invariant and $\beta P^1 \beta(e_1U \otimes 1)$.

Now comes the clincher. We consider the Atiyah-Hirzebruch spectral sequences for computing $\pi_*(BP \wedge K(\mathbb{Z}/p,2))$ and $\pi_*(M\,\mathrm{Cok}\,J \wedge BP \wedge K(\mathbb{Z}/p,2))$. The E_2-term in the first case is $H_*(K(\mathbb{Z}/p,2)),\ \pi_*(BP)) \simeq \pi_*(K(\pi_*(BP)) \wedge K(\mathbb{Z}/p,2))$. The differentials are exactly described by the k-invariants of BP (which were actually all written down in Brown and Peterson's original paper on BP). The pattern of differentials might be tedious to compute from scratch, but we already have a lot of information on $\pi_*(BP \wedge K(Z/p,2))$ by our Adams spectral sequence method, and, we know the Atiyah-Hirzebruch differentials commute with the action of $\pi_*(BP)$ on $H_*(K(Z/p,2), \pi_*(BP))$. It is then easy to identify the only possible surviving terms of filtration greater than zero in E_∞, which is the Hurewicz kernel. These are groups Z/p which are quotients of groups Z/p^2 in E_2 in terms $H_{2ip}(K(Z/p,2), \pi_{4j}(BP))$. But we know that $M\,Cok\,J \wedge BP$ has the same k-invariants as BP except for a perturbation in high filtration. Thus the differentials on $H_*(K(Z/p,2), \pi_*(M\,Cok\,J \wedge BP))$ will agree with those on $H_*(K(Z/p,2), \pi_*(BP))$ until this high filtration level is reached. It is then simple to just write down the two E_2 terms and observe that the survivors to E_∞ in the case $\pi_*(BP \wedge K(Z/p,2))$ also survive to E_∞ in $\pi_*(M\,Cok\,J \wedge BP \wedge K(Z/p,2))$. This is our long sought after injectivity $\pi_*(BP \wedge K(Z/p,2)) \to \pi_*(M\,Cok\,J \wedge BP \wedge K(Z/p,2))$.

REFERENCES

1. E. H. Brown Jr. and F. P. Peterson, A spectrum whose Z/p cohomology is the algebra of reduced p^{th} powers, Topology 5 (1966), 149-154.

2. S. Gitler and J. Stasheff, The first exotic class of BF, Topology 4 (1965), 227-266.

3. F. P. Peterson, Some results on PL cobordism, J. Math. Kyoto Univ. 9 (1969), 189-194.

4. R. Thom, Quelques propriétés globale des variétés differentiable, Comm. Math. Helv. 28 (1954). 17-86.

TOPOLOGY AND MEASURE

by Leon W. Cohen

University of Maryland

The interaction of topology and measure in providing a basis for analysis is evident in the pioneer work of Borel and Lebesgue. Here this interaction is exhibited in the properties of intersections

$$G \cap C$$

where G is open and C is compact in a topological space. These sets imitate the half-open intervals

$$I^n = \mathop{X}_{k=1}^{n} (a_k, b_k] \subset R^n$$

very closely. Using this analogy it is shown that if S is a metric space or, more generally, a perfectly normal topological space (i.e., one in which each open set G is the union of a countable set of closed sets F_m) a useful measure can be defined. This serves as a model for a theory of measure in an abstract space S which combines the ideas of Borel with Lebesgue's exterior measure.

As an application some indications are given as to how the class of almost continuous functions can substitute for the class of measurable functions in analysis.

Proposition 1. If S is a topological space and

$$\mathcal{J} = \{G \cap C \mid G \text{ open and } C \text{ compact in } S\}$$

than the empty set $\emptyset \in \mathcal{J}$ and for $I, J \in \mathcal{J}$

(i_1) $I \cap J \in \mathcal{J}$

(i_2) $I - J = I_1 \cup I_2 \cup I_3$ where I_1, I_2, I_3 are disjoint and in $\mathcal{J}$.

Remark. The I^n in R^n have these properties except that in (i_2) 3 is replaced by $m \leq 3^n - 1$.

Proof. Details are unnecessary except for (i_2). Consider $I = G \cap C$, $J = H \cap K$

where G, H are open and C, K are compact. Then

$$I - J = G \cap C \cap \{S - H \cap K\} = G \cap C \cap \{(S - H) \cup (S - K)\}$$

where $S - H = F$ is closed, $S - K = G'$ is open. Now

$$(S - H) \cup (S - K) = (F - G') \cup (F \cap G') \cup (G' - F)$$

where $F - G' = F'$ is closed, $G' - F = G''$ is open and $F', (FG'), G''$ are disjoint. Then $I_1 = G \cap (C \cap F')$, $I_2 = (G \cap G') \cap (C \cap F')$, $I_3 = (G \cap G'') \cap C$ are disjoint sets in $\mathcal{J}$ whose union is $I - J$.

We consider an abstract space S, to be called a <u>space</u>. An <u>interval structure</u> in S is a non-empty set $\mathcal{J}$ of subsets I of S, to be called <u>intervals</u> in S such that for any $I, J \in \mathcal{J}$

(i_1) $I \cap J \in \mathcal{J}$

(i_2) $I - J = \bigcup_{k=1}^{m} I_k$ for some disjoint intervals $I_1, \ldots, I_m$

If S is a topological space the set

$$\mathcal{J} = \{G \cap C \mid G \text{ open and } C \text{ compact in } S\}$$

which, since $\mathcal{J}$ contains $\emptyset$ and, by Proposition 1, satisfies (i_1), (i_2), is an interval structure in S. Since this is unique, it is the <u>topological interval</u> structure in S. The $I \in \mathcal{J}$ are the <u>topological intervals</u> in S.

Given a space S and an interval structure $\mathcal{J}$ in S, an <u>interval measure</u> on $\mathcal{J}$ is a function λ on $\mathcal{J}$ to R^+ such that

(λ) $$\lambda(\emptyset) = 0$$

Considering R^n as the familiar metric space the Lebesgue measure is an interval measure on the topological interval structure in R^n.

Given S, $\mathcal{J}$ in S and λ on $\mathcal{J}$ satisfying (i_1), (i_2), (λ), we define the auxiliary set

$$\mathcal{S} = \{A \subset S \mid A \subset \bigcup_n I_n \text{ for some sequence } I_n \text{ in } \mathcal{J}\}$$

and the function μ on $\mathcal{S}$ to R^+_∞ where for $A \in \mathcal{J}$

$$\mu(A) = \inf\left\{\sum_n \lambda(I_n) \mid A \subset \bigcup_n I_n \text{ and } I_n \in \mathcal{J} \text{ for all } n\right\}.$$

We note that

(a) $\mathcal{S}$ is a complete ring, i.e.

$A - B \in \mathcal{S}$ if $A, B \in \mathcal{S}$, $\bigcap_n A_n$ and $\bigcup_n A_n \in \mathcal{S}$ for any sequence of $A_n \in \mathcal{S}$.

(b) μ is subadditive on $\mathcal{S}$, i.e.

$$\mu(\bigcup_n A_n) \leq \sum_n \mu(A_n) \quad \text{for all sequences of } A_n \in \mathcal{S}.$$

(c) The set

$$\mathcal{N} = \{A \in \mid \mu(A) = 0\}$$

is a complete subring of $\mathcal{S}$.

Given a space S, an interval structure $\mathcal{J}$ in S and an interval measure λ on $\mathcal{J}$, the pair $\{\mathcal{J},\lambda\}$ is a <u>measure structure</u> in S and μ is the <u>associated measure</u> if

(μ) For any disjoint intervals $I_1, \ldots, I_n$,

$$\mu(\bigcup_{k=1}^{n} I_k) = \sum_{k=1} \lambda(I_k).$$

If S is a topological space and $\mathcal{J}$ is the topological measure structure in S, a measure structure $\{\mathcal{J},\lambda\}$ is a <u>topological measure structure</u>. It is not unique since the interval measure λ is not unique.

The proof of the next proposition is elementary.

<u>Proposition 2</u>. <u>If</u> μ <u>is the measure associated with a measure structure</u> $\{\mathcal{J},\lambda\}$ <u>in</u>

S .

(a) $\mu(I) = \lambda(I)$ *for all* $I \in \mathcal{J}$.

(b) *If* $I, I_1, \ldots, I_n$ *are intervals there are disjoint intervals* $J_1, \ldots, J_m$ *such that*

$$I - \bigcup_{k=1}^{m} I_k = \bigcup_{k=1}^{m} J_k .$$

(c) *If* $I_n \in \mathcal{J}$ *for* $n \in N$ *there are disjoint* $J_n \in \mathcal{J}$ *for* $n \in N$ *such that*

$$\bigcup_n I_n = \bigcup_n J_n .$$

and for all sequences of disjoint intervals J_n *satisfying* (c)

(d) $$\mu(\bigcup_n I_n) = \sum_n \mu(J_n) .$$

The measure μ is, of course, suggested by Lebesgue's exterior measure. The set

$$\mathcal{J}_\sigma = \{\bigcup_n I_n \mid I_n \in \mathcal{J} \text{ for all } n\}$$

together with Proposition 2 are suggested by Borel. From here on we have been guided by Borel, in particular by considering the set $\mathcal{J}_\sigma$ in detail. It is not the case, if $S = \mathbf{R}$, $\mathcal{S} = \{(a,b] \mid a \leq b \text{ in } \mathbf{R}\}$ and $\lambda((a,b]) = b - a$, that $A, B \in \mathcal{J}_\sigma$ implies $B - A \in \mathcal{J}_\sigma$ although $\{\mathcal{J}, \lambda\}$ is a measure structure in $\mathbf{R}$.

The next proposition shows that in a class of topological spaces, which includes metric space, the defect referred to just above does not occur.

Proposition 3. *If* S *is a topological space such that for each open set* G *there is a sequence* F_n *of closed sets such that*

$$G = \bigcup_n F_n$$

(such spaces are perfectly normal) and I_n, J_n *are sequences of topological inter-*

<u>vals in</u> S <u>then there is a sequence of topological intervals</u> I'_n <u>such that</u>

$$\bigcup_n I_n - \bigcup_n J_n = \bigcup_n I'_n$$

<u>Proof</u>. There are open sets G_n, H_n and compact sets C_n, K_n for $n \in N$ such that

(1) $$\bigcup_n I_n - \bigcup_n J_n = \bigcup_n (G_n \cap C_n) - \bigcup_n (H_n \cap K_n)$$

$$= \bigcap_m [\bigcup_n (G_n \cap C_n) \cap \{(S - H_m) \cup (S - K_m)\}].$$

Since for each m, $S - K_m$ is open, it follows from the hypothesis that there are closed sets $F_{m,s}$, $s \in N$, such that

$$S - K_m = \bigcup_s F_{m,s} .$$

Since $S - H_m = F_m$ is closed, $F_m \cup F_{m,s} = F'_{m,s}$ is closed for all m, s. Now

$$(S - H_m) \cup (S - K_m) = \bigcup_s F'_{m,s} \quad \text{for all } m$$

and, from (1),

(2) $$\bigcup_n I_n - \bigcup_n J_n = \bigcup_{n,s} (G_n \cap (C_n \bigcap_m F'_{m,s})) .$$

Since C_n is compact and $F'_{m,s}$ is closed for all n, m, s

$$C_{n,s} = C_n \cap (\bigcap_m F'_{m,s}) \text{ is compact for all } n, s .$$

Now $\{I_{n,s} = G_n \cap C_{n,s} | n,s \in N\}$ is a countable set of topological intervals I'_n and

$$\bigcup_n I_n - \bigcup_n J_n = \bigcup_n I'_n$$

follows from (2).

<u>Proposition 4</u>. <u>If</u> $\mathcal{J}$ <u>is an interval structure in a space</u> S <u>and</u> $A, B \in \mathcal{J}_\sigma =$

$\{\bigcup_n I_n \mid I_n \in \mathcal{J}$ <u>for all</u> $n\}$ <u>implies</u> $A - B \in \mathcal{J}_\sigma$, <u>then</u> $\mathcal{J}_\sigma$ <u>is a complete ring</u>.

<u>Proof</u>. Consider any sequence of sets $A_n \in \mathcal{J}_\sigma$. For each n there is a sequence of intervals $\mathcal{J}_{n,m}$ such that

$$A_n = \bigcup_m I_{n,m} .$$

It is evident that

$$\bigcup_n A_n = \bigcup_{n,m} I_{n,m} \in \mathcal{J}_\sigma . \tag{1}$$

For each $p \in N$

$$\bigcap_{n=1}^{p} A_n = \bigcap_{n=1}^{p} \bigcup_m I_{n,m} = \bigcup_{m_1,\ldots,m_p \in N} (I_{1,m_1} \cap I_{2,m_2} \cap \ldots \cap I_{p,m_p}) .$$

Since from (i_1) and induction

$$I_{m_1,\ldots,m_p} = \bigcap_{q=1}^{p} I_{q,mq} \in \mathcal{J}$$

and $\{I_{m_1,\ldots,m_p} \mid m_1,\ldots,m_p \in N\}$ is a countable subset of $\mathcal{J}$.

$$\bigcap_{n=1}^{p} A_n \in \mathcal{J}_\sigma \quad \text{for all } p . \tag{2}$$

By the hypothesis, (1), and (2)

$$\bigcap_n A_n = A_1 - \bigcup_p \left(\bigcap_{n=1}^{p} A_n - \bigcap_{n=1}^{p+1} A_n\right) \in \mathcal{J}_\sigma . \tag{3}$$

From the hypothesis, (1) and (3) it follows that $\mathcal{J}_\sigma$ is a complete ring.

<u>Remark</u>. The condition

(c) If $A, B \in \mathcal{J}_\sigma$ then $A - B \in \mathcal{J}_\sigma$

is a completeness condition on $\mathcal{J}_\sigma$ in much the same way that the Cauchy condition is a completeness condition on $\mathcal{R}$ - it assures the existence in $\mathcal{J}_\sigma$ of the limit of any convergent sequence of sets in $\mathcal{J}_\sigma$.

We recall some definition. If A_n is any sequence of sets,

$$\liminf_n A_n = \bigcup_p \bigcap_{n \geq p} A_n , \quad \limsup_n A_n = \bigcap_p \bigcup_{n \geq p} A_n .$$

The sequence <u>converges</u> to $A = \lim_n A_n$ if

$$\liminf_n A_n = A = \limsup_n A_n .$$

A set $\mathcal{C}$ of sets is <u>complete</u> if $\mathcal{C}$ contains $\lim_n A_n$ for any convergent sequence of $A_n \in \mathcal{C}$.

From Proposition 3 it follows that if $\mathcal{J}$ is the topological interval structure in a perfectly normal space, then $\mathcal{J}_\sigma = \{ \bigcup_n I_n | I_n \in \mathcal{J} \text{ for all } n\}$ is a complete ring and consequently complete as a set of sets.

A measure structure $\{\mathcal{J},\lambda\}$ in a space S is a <u>complete measure structure</u> if $\mathcal{J}_\sigma$ satisfies the completeness condition

(c) If $A , B \in \mathcal{J}_\sigma$ then $A - B \in \mathcal{J}_\sigma$.

Now we state our main result:

<u>Theorem</u>. *If $\{\mathcal{J},\lambda\}$ is a complete measure structure in a space S and μ is the associated measure then*

$$\mathfrak{M} = \{A \in \mathcal{S} | \mu(A - B) = \mu(B - A) = 0 \text{ for some } B \in \mathcal{J}_\sigma\}$$

is a complete ring. If $A_n \in \mathfrak{M}$ for $n \in N$ then

(Λ) *if A_n converges and $\mu(\bigcup_n A_n) < \infty$, $\mu(\lim_n A_n) = \lim_n \mu(A_n)$,*

(Σ) *if the A_n are disjoint then $\mu(\bigcup_n A_n) = \sum_n \mu(A_n)$.*

The next propositions enter the proof.

<u>Proposition 5</u>. *If μ is the measure associated with a measure structure $\{\mathcal{J},\lambda\}$ in*

<u>a space</u> S , A , B $\in \mathcal{J}_\sigma$ <u>and</u> $A \subset B$ <u>then</u>

$$\mu(B) = \mu(A) + \mu(B - A)$$

[Note: B - A need not be in $\mathcal{J}_\sigma$.]

<u>Proof</u>. Since μ is subadditive on the complete ring $\mathcal{S}$

(1) $$\mu(B) \leq \mu(A) + \mu(B - A).$$

The equality holds if $\mu(B) = \infty$. Assume $\mu(B) < \infty$. There are sequences of intervals I_n, J_n such that

$$\bigcup_n I_n = A \subset B = \bigcup_n J_n$$

By Proposition 2(d) no generality is lost by assuming that the I_n and the J_n are disjoint. Now the intervals $I_n \cap J_m$ are disjoint for $n, m \in N$ and

$$\mu(A) = \sum_{n,m} \mu(I_n \cap J_m) \leq \mu(B) = \sum_m \mu(J_m) < \infty.$$

For all $k \in N$

$$B - A = \bigcup_m J_m - \bigcup_m \bigcup_n (I_n \cap J_m) \subset \bigcup_m \{J_m - \bigcup_{n=1}^{k} (I_n \cap J_m)\},$$

$$\mu(B - A) \leq \sum_m \mu\{J_m - \bigcup_{n=1}^{k} (I_n \cap J_m)\}.$$

By Proposition 2(b) there are disjoint intervals $I_{m,k,s}$, $s = 1, \ldots, t_{m,k} = t$, such that

$$J_m - \bigcup_{k=1}^{n} (I_n \cap J_m) = \bigcup_{s=1}^{t} I_{m,k,s}$$

From (μ) it follows that

$$\mu(J_m) = \sum_{n=1}^{k} \mu(I_n \cap J_m) + \sum_{s=1}^{t} \mu(I_{m,k,s})$$

$$= \sum_{n=1}^{k} \mu(I_n \cap J_n) + \mu\{J_m - \bigcup_{n=1}^{k} (I_n \cap J_m)\} \quad \text{for all} \quad m, k .$$

For $\epsilon > 0$ there is some $k = k_\epsilon$ such that

$$\mu(A) + \mu(B - A) \leq \sum_m \sum_{n=1}^{k} \mu(I_n \cap J_m) + \epsilon + \sum_m \mu\{J_m - \sum_{n=1}^{k} (I_n \cap J_m)\}$$

$$= \sum_m \mu(J_m) + \epsilon = \mu(B) + \epsilon .$$

The required equality follows from (1) and (2).

<u>Proposition 6</u>. *If* μ *is the measure associated with a measure structure* $\{\mathcal{J}, \lambda\}$ *in a space* S *and* $A_n \in \mathcal{J}_\sigma$ *for* $n \in N$, *then*

(a) $\mu(\bigcup_n A_n) = \lim_n \mu(A_n)$ *if* $A_n \subset A_{n+1}$ *for all* n,

(b) $\mu(\bigcap_n A_n) = \lim_n \mu(A_n)$ *if* $A_n \supset A_{n+1}$ *for all* n *and* $\mu(A_1) < \infty$,

(c) $\mu(\bigcup_n A_n) = \sum_n \mu(A_n)$ *if the* A_n *are disjoint*.

<u>Proof</u>. Assume that $A_n \subset A_{n+1}$ for all n. Then

$$A_n \subset A_{n+1} \subset \bigcup_n A_n = A_1 \cap \bigcup_n (A_{n+1} - A_n) ,$$

$$\mu(A_n) \leq \mu(A_{n+1}) \leq \sup_n \mu(A_n) \leq \mu(\bigcup_n A_n) \leq \mu(A_1) + \sum_n \mu(A_{n+1} - A_n) \leq \infty .$$

If $\sup_n \mu(A_n) = \infty$, (a) holds. Assume $\sup_n \mu(A_n) < \infty$. Then by Proposition 4

$$\mu(A_{n+1} - A_n) = \mu(A_{n+1}) - \mu(A_n) \quad \text{for all} \quad n$$

and

$$\lim_n \mu(A_n) \leq \mu(\bigcup_n A_n) \leq \mu(A_1) + \sum_n (\mu(A_{n+1}) - \mu(A_n) = \lim_n \mu(A_n) .$$

This proves (a).

Assume $A_n \supset A_{n+1}$ for all n and $\mu(A_1) < \infty$. Then

$$\bigcap_n A_n \subset A_{n+1} \subset A_n \subset A_1 \,, \quad A_1 = \bigcap_n A_n \cup \bigcup_n (A_n - A_{n+1}) \,,$$

(1) $$\mu(\bigcap_n A_n) \leq \inf_n \mu(A_n) \leq \mu(A_{n+1}) \leq \mu(A_n) < \infty \quad \text{for all } n \,,$$

$$\mu(A_1) \leq \mu(\bigcap_n A_n) + \sum_n \mu(A_n - A_{n+1}) \,.$$

By Proposition 4

$$\mu(A_n - A_{n+1}) = \mu(A_n) - \mu(A_{n+1}) \,.$$

Hence

(2) $$\mu(A_1) \leq (\bigcap_n A_n) + \sum_n (\mu(A_n) - \mu(A_{n+1})) = \mu(A_1) - \lim_n \mu(A_n)$$

and, from (1), (2),

$$\lim_n \mu(A_n) \leq \mu(\bigcap_n A_n) \leq \lim_n \mu(A_n) \,.$$

which yields (b).

Assume that the A_n are disjoint. From Proposition 2(c), (d) it follows that for each n there are disjoint intervals $I_{n,m}$ for $m \in N$ such that

$$A_n = \bigcup_m I_{n,m} \,, \quad \mu(A_n) = \sum_m \mu(I_{n,m}) \,,$$

$$A_n = \bigcup_{n,m} I_{n,m} \,,$$

$$\mu(\bigcup_n A_n) = \sum_{n,m} \mu(I_{n,m}) = \sum_n \sum_m \mu(I_{n,m}) = \sum_n \mu(A_n)$$

since $\{I_{n,m} \mid n,m \in N\}$ is a countable set of disjoint intervals.

<u>Proposition 7</u>. <u>If μ is the measure associated with a complete measure structure $\{\mathcal{J},\lambda\}$ in S and A_n is a sequence of sets in $\mathcal{J}_\sigma$ then</u>

(a) $\mu(\bigcup_n A_n) = \lim_n \mu(A_n)$ <u>if</u> $A_n \subset A_{n+1}$ <u>for all</u> n,

(b) $\mu(\bigcap_n A_n) = \lim_n \mu(A_n)$ <u>if</u> $A_n \supset A_{n+1}$ <u>for all</u> n, <u>and</u> $\mu(A_1) < \infty$,

(c) $\mu(\bigcup_n A_n) = \sum_n \mu(A_n)$ <u>if the</u> A_n <u>are disjoint</u>,

(d) $\mu(\liminf_n A_n) \leq \liminf_n \mu(A_n) \leq \limsup_n \mu(A_n) \leq \mu(\limsup_n A_n) < \infty$

<u>if</u> $\mu(\bigcup_n A_n) < \infty$

<u>and all sets which occur are in</u> $\mathcal{J}_\sigma$.

<u>Proof</u>. Since $\{\mathcal{J},\lambda\}$ is a complete measure structure it follows from Proposition 4 that $\mathcal{J}_\sigma$ is a complete ring and since $A_n \in \mathcal{J}_\sigma$ for all n

$$\bigcup_n A_n\,,\ \bigcap_n A_n\,,\ \bigcup_p \bigcap_{n\geq p} A_n\,,\ \bigcap_p \bigcup_{n\geq p} A_n \in \mathcal{J}_\sigma .$$

Parts (a), (b), (c) have been established by Proposition 6.

For (d) note that for $p \in N$

(1) $$\bigcap_{n\geq p-1} A_n \subset \bigcap_{n\geq p} A_n \subset A_n \quad \text{for } n > p\,,$$

(2) $$A_n \subset \bigcup_{n\geq p} A_n \subset \bigcup_{n\geq p-1} A_n \subset \bigcup_n A_n \quad \text{for } n > p\,.$$

Hence

(3) $$\mu(\bigcap_{n\geq p-1} A_n) \leq \mu(\bigcap_{n\geq p} A_n) \leq \inf_{n\geq p} \mu(A_n) \quad \text{for all } p\,,$$

(4) $$\sup_{n>p} \mu(A_n) \leq \mu(\bigcup_{n\geq p} A_n) \leq \mu(\bigcup_{n\geq p-1} A_n) \leq \mu(\bigcup_n A_n) \quad \text{for all } p\,.$$

From (a), (1), (3) follows

(5) $$\mu(\liminf_n A_n) = \mu(\bigcup_p \bigcap_{n\geq p} A_n) = \sup_p \mu(\bigcap_{n\geq p} A_n)$$

$$\sup_p \inf_{n>p} \mu(A_n) = \liminf_n \mu(A_n)\,.$$

From (b), (2), (4) follows

(6) $$\limsup_n \mu(A_n) = \inf_p \sup_{n>p} \mu(A_n) \geq \inf \mu(\bigcup_{n\geq p} A_n)$$

$$= \mu(\bigcap_p \bigcup_{n\geq p} A_n) = \mu(\limsup_n A_n) < \infty .$$

Since $\mu(A_n)$ is a sequence of real numbers, (d) follows from (5), (6).

Corollary. *If A_n is a convergent sequence in $\mathcal{J}_\sigma$ then*

$$\lim_n \mu(A_n) = \mu(\lim_n A_n)$$

provided A_n is an increasing sequence and in all other cases provided $\mu(\bigcup_n A_n) < \infty$.

We now consider the set of μ-*measurable* subsets of S,

$$\mathfrak{M} = \{A \in \mathcal{S} \mid \mu(A - B) = \mu(B - A) = 0 \text{ for some } B \in \mathcal{J}_\sigma\}$$

under the assumption that μ is the measure associated with a complete measure structure $\{\mathcal{J},\lambda\}$ in S. We recall that $\mathcal{J}$, λ, μ satisfy the conditions (i_1), (i_2), (c), (λ), (μ).

Let us define: A is μ-*equivalent* to B in $\mathcal{S}$, $A \underset{\mu}{\sim} B$, if $A, B \in \mathcal{S}$ and

$$\mu(A - B) = \mu(B - A) = 0$$

It is evident that

Proposition 8. *$A \underset{\mu}{\sim} B$ is an equivalence relation in $\mathcal{S}$.*

If $A \underset{\mu}{\sim} B$ then $\mu(A) = \mu(B)$.

If $A' \underset{\mu}{\sim} B'$ and $A'' \underset{\mu}{\sim} B''$ then $A' - A'' \underset{\mu}{\sim} B' - B''$.

If $A_n \underset{\mu}{\sim} B_n$ for $n \in N$ then $\bigcup_n A_n \underset{\mu}{\sim} \bigcup_n B_n$ and $\bigcap_n A_n \sim \bigcap_n B_n$.

It is evident that

$$\mathfrak{M} = \{A \in \mathcal{S} \mid A \underset{\mu}{\sim} B \text{ for some } B \in \mathcal{J}_\sigma\}$$

and that

$$\mathcal{J} \subset \mathcal{J}_\sigma \subset \mathcal{J}_\sigma \cup \mathcal{N} \subset \mathcal{M} \subset \mathcal{S}.$$

Theorem. If μ is the measure associated with a complete measure structure on a space S, the set $\mathcal{M}$ of μ-measurable sets is a complete ring on which μ has the properties of Proposition 7.

Proof. That $\mathcal{M}$ is a complete ring follows from Proposition 8 since $\mathcal{J}_\sigma$ is a complete ring. Properties (a), (b), (d) follow quickly from Propositions 7, 8.

For (c), consider a sequence of disjoint A_n in $\mathcal{M}$. For each n there is some $B_n \in \mathcal{J}_\sigma$ such that $A_n \underset{\mu}{\sim} B_n$. Now

$$\mu(A_n) = \mu(B_n) \quad \text{for all} \quad n,$$

and

$$\mu(\bigcup_n B_n) = B_1 \cup \{\bigcup_n (B_{n+1} - B_{n+1} \cap (\bigcup_{k=1}^{n} B_k))\}$$

where

$$B_1,\ B_{n+1} - (\bigcup_{k=1}^{n} B_k),\quad n \in N$$

are in $\mathcal{J}_\sigma$ and are disjoint. From Proposition 8 and the disjointness of the A_n,

$$B_{n+1} - B_{n+1} \cap (\bigcup_{k=1}^{n} B_n) \underset{\mu}{\sim} A_{n+1} - A_{n+1} \cap (\bigcup_{k=1}^{n} A_k) = A_{n+1}.$$

Hence

$$\mu(\bigcup_n A_n) = \mu(\bigcup_n B_n) = \mu(B_1) + \sum_n \mu(B_{n+1} - B_{n+1} \cap (\bigcup_{k=1}^{n} k) = \sum_n \mu(A_n) .$$

As an application, consider a metric space S with a topological measure structure $\{\mathcal{J},\lambda\}$ and the associated measure μ. The concept of continuity may be used as a substitute for measurability in the theory of real functions in S.

Since S is open the compact sets are topological intervals.

Definition. A (real) function f is almost continuous on its domain $A \in \mathcal{M}$ if

<u>for</u> $\epsilon > 0$ <u>there is a compact</u> $C \subset A$ <u>such that</u>

$$f \text{ is continuous on } C \text{ and } \mu(A - C) < \epsilon.$$

It is shown [Cohen] that

1. If $A \in \mathfrak{M}$ and $\mu(A) < \infty$ there is a compact $C \subset A$ such that $\mu(A - C) < \epsilon$.
2. If f_n is a convergent sequence of functions continuous on a compact C and $\epsilon > 0$ there is a compact $B \subset C$ such that $\mu(C - B) < \epsilon$ and the f_n converge uniformly on B. (This is a form of Egorov's theorem [Egorov].)
3. If f_n is a sequence of functions almost continuous on $A \in \mathfrak{M}$ and $\lim_n f_n = f$ on A then f is almost continuous on A.
4. If f is continuous on a compact set C then

$$\int_C f d\mu = \lim_{\pi} \sum_{k=1}^{n} f(x_k)\mu(I_k) \in \mathbb{R}$$

where $\pi = \{I_1, \ldots, I_k\}$ is a partition of C into disjoint intervals and $x_k \in I_k$.

The proof is essentially that of Cauchy.

<u>Definition</u>. <u>The integral of an almost continuous function</u> f <u>over</u> $A \in \mathfrak{M}$ <u>is</u>

$$\int_A f d\mu = \alpha$$

<u>if, for each sequence of compact sets</u> $C_n \subset A$ <u>such that</u> f <u>is continuous on</u> C_n <u>and</u> $\lim_n \mu(A - C_n) = 0$,

$$\lim_n \int_{C_n} fd = \alpha .$$

5. The interchange formula

$$\int_A \lim_n f_n \, d\mu = \lim_n \int_A f_n \, d\mu$$

holds under the conditions for Lebesgue's monotonic convergence theorem and the conditions of the dominated convergence theorem. The proof in the latter case makes use of the uniformity in 2 .

HISTORICAL REMARKS

The beginnings of the current theory of real functions lie largely in two works:

Borel, E., Leçons sur la theorie des fonctions, Paris, 1898 where he may have had in mind, but did not state, something like the completeness condition for the interval structure in $\mathbb{R}$.

Lebesgue, H., Leçons sur l'integration et la recherche des fonctions primitives, Paris, 1904 where he introduced the concept of exterior measure and the concept of measurable set, closing the gap in Borel's work.

The definition of almost continuity is taken from Lusin's theorem which essentially states that a measurable function is almost continuous.

Lusin, N., Sur les proprietes des fonctions measurables. C.R. Acad. Sci. 154 (1912), 1688-1690.

The theorem of Egorov is more general, using measurable sets instead of compact sets and measurable functions instead of continuous functions.

Egorov, D.T., Sur les suites des fonctions measurables, C.R. Acad. Sci. 152 (1911), 244-246.

The details for the statements made as applications are found in

Cohen, L.W., Measure and integration in the manner of Borel, Scripta Math. XXIX (1972), 417-435.

The reader should avoid reading pages 417-423 in which a defective account of the content of this paper is found.

$\mathbb{H}P^\infty$, Genuine and Counterfeit

Morton Curtis and George Terrell

Rice University

I. Introduction

1. Classifying spaces

If G is a topological group, there is a construction due to Milnor [7] of a principal G-bundle $G \to E \to B_G$ with contractible total space E. Thus $G \to E \to B_G$ is a universal G-bundle and B_G is the classifying space. The assignment $G \to B_G$ is functional: A homomorphism $\varphi: G \to H$ yields a map $B(\varphi): B_G \to B_H$.

EXAMPLES:

$$G = \mathbb{Z}/2 \qquad B_G = RP^\infty$$
$$G = S^1 \qquad B_G = CP^\infty$$
$$G = Sp(1) \qquad B_G = \mathbb{H}P^\infty$$

($Sp(1) \cong SU(2)$ is the 3-sphere.)

Consider the Hopf fibration $S^1 \overset{f}{\to} S^3 \to S^2$. This is induced by a map $S^2 \to B_{S^1} = CP^\infty$. We have

$$\begin{array}{ccccc} S^1 & = & S^1 & & \\ \downarrow & & \downarrow & & \\ S^3 & \to & E & \text{(contractible)} & \\ \downarrow & & \downarrow & & \\ S^2 & \to & CP^\infty & \overset{B(f)}{\to} & \mathbb{H}P^\infty \end{array}$$

and

$$S^1 \overset{f}{\to} S^3 \to S^2 \to CP^\infty \overset{B(f)}{\to} \mathbb{H}P^\infty$$

is part of the "fiber sequence" (see e.g. [8] page 138) of the map $B(f)$.

2. Localization, mixes

We denote the integers localized at a prime p by

$$\mathbb{Z}_p = \{\tfrac{a}{b} \mid p \nmid b\} \subset \mathbb{Q}.$$

Thus every prime $q \neq p$ is a unit in $\mathbb{Z}_p$ because $\frac{1}{q} \in \mathbb{Z}_p$. If we also make p into a unit we get $Z_p \to Z_0 = Q$. Now if G is an abelian group we define G localized at p to be $G \otimes Z_p$.

A space X localized at p is a space X_p and a map $f: X \to X_p$ such that

$$f_\# \pi_i(X) = \pi_i(X_p) \cong (\pi_i(X))_p.$$

With some reasonable conditions on the space X ([5],[11]) we get a diagram

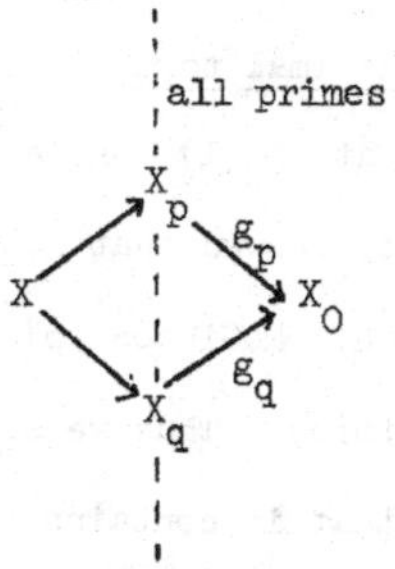

and X may be obtained from the fiber product of this diagram by killing the $\lim^1$ terms in π_3. A <u>mix</u> is obtained by modifying some (or all) of the maps g_p to get a new "fiber product" X'.

3. Rector's examples

There are some especially interesting examples due to Rector [10]. Let $X = \mathbb{H}P^\infty$. We have

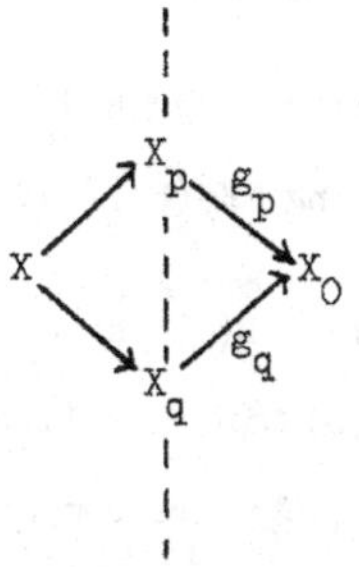

and X_0 is the Eilenberg-MacLane space $K(Q,4)$. Thus g_p is a rational cohomology class of X_p and if we choose $r_p \in Q$, $r_p g_p$ is a new map of X_p into X_0. For choices $\{r_p\}$ we denote the fiber product by $X\{r_p\}$. Rector showed that this yields an uncountable infinitude $\{X_\alpha\}$ of homotopy types, each with $\Omega X_\alpha \cong S^3$. With each $r_p = 1$ we get the <u>genuine</u> $\mathbb{H}P^\infty$ and the other homotopy types are called <u>fake</u> (or counterfeit) $\mathbb{H}P^\infty$'s.

Two spaces X,Y have the same <u>genus</u> if $X_p \simeq Y_p$ for all primes p. These examples constitute the genus of $\mathbb{H}P^\infty$.

4. <u>Maximal Tori</u>

Let X be a space such that $\Omega X \simeq$ conn. finite CW. Then Hopf showed $H^*(\Omega X;Q) = \Lambda(x_1,\ldots,x_r)$ and we call r the <u>rank</u> of ΩX. Let $Y = \underbrace{CP^\infty \times \cdots \times CP^\infty}_{r}$. Rector [9] says that ΩX has a <u>maximal torus</u> if there exists a fibration $F \to Y \to X$ with $F \simeq$ finite CW. We know that $S_p(1)$ has a maximal torus from the fibration $S^2 \to CP^\infty \to \mathbb{H}P^\infty$ (see §1). Rector showed that none of the fakes do have a maximal torus, with one possible exception. McGibbon [6] has recently proved that this fake also fails to contain a maximal torus. Thus we now know that the genuine $\mathbb{H}P^\infty$ is distinguished from all fakes in that it contains a maximal torus. We discuss here another approach (due to the second author). Proofs will appear later.

II. Complex K-theory of $\mathbb{H}P^\infty$

1. <u>Representation rings</u>

It is a classical theorem of Atiyah (see [2]) that if G is a compact connected Lie group with classifying space B_G, then the two rings

(i) the complex representation ring $R(G)$ of G

(ii) the complex K-theory ring $K^*(B_G)$ of B_G

are closely related. In particular $R(G)$ sits naturally in $K^*(B_G)$ and $K^*(B_G)$ is just the completion $\widehat{R(G)}$ of $R(G)$ in the filtration topology on $K^*(B_G)$.

This is "functorial": If G_1, G_2 are compact connected Lie groups and $f: B_{G_1} \to B_{G_2}$ is any map, then

$$f^!: K^*(B_{G_2}) \to K^*(B_{G_1})$$

sends $R(G_2)$ into $R(G_1)$. (See [1] Corollary 1.13).

DEFINITION: Let R be a formal power series ring. A subring P of R consisting of polynomials is said to be full in R if P has a set of generators which is also a set of generators for R.

Again, let $X = \mathbb{H}P^{\infty} = BS_p(1)$. From the results of Atiyah cited above we see that:

$$R(S_p(1)) \quad \text{is full in} \quad K^*(X).$$

CONJECTURE: If X_{α} is a fake $\mathbb{H}P^{\infty}$, then $K^*(X_{\alpha})$ has no full polynomial subring.

2. Calculating $f^!$ of a map $X \xrightarrow{f} X$.

$H^*(X)$ is a polynomial ring on a 4-dimensional generator and the degree of $f: X \to X$ is the degree on such a generator. Berstein [3] has shown that generators for H^* and K^* of CP^{∞} and $\mathbb{H}P^{\infty}$ can be chosen as follows. Let $\alpha: CP^{\infty} \to \mathbb{H}P^{\infty}$ be the base space map coming from the Hopf map $S^1 \to S_p(1)$. Let c be the chern class. We have

$$\begin{array}{ccccc} \xi & K^*(\mathbb{H}P^{\infty}) & \xrightarrow{\alpha^!} & K^*(CP^{\infty}) & \eta \\ & \Big\downarrow c & & \Big\downarrow c & \\ y & H^*(\mathbb{H}P^{\infty}) & \xrightarrow{\alpha^*} & H^*(CP^{\infty}) & t \end{array}$$

such that $c(\xi) = 1 - y$, $c(\eta) = 1 + t$ and

$$\alpha^*(y) = t^2.$$

THEOREM 1: Set $x = \xi - 2$. If $f: X \to X$ has a degree a (i.e. $f^*y = ay$), then

$$(\chi) \qquad f^!(x) = \sum_{n=1}^{\infty} \left\{ \frac{2}{(2n)!} \prod_{i=0}^{n-1} (a - i^2) \right\} x^n$$

COROLLARY (Bernstein): a is a square.

PROOF: Replace x by the real number 1. The series converges (ratio test) and coefficients are integers. Thus a is some i^2.

Berstein has shown that the Adams operation ψ^p has the same effect on $K^*(X)$

as a map of degree p^2 (and Sullivan [11, Cor. 5.10] has shown that such maps do exist for p odd). So we have

$$(*) \qquad \psi^p(X) = \sum_{n=1}^{\infty}\left\{\frac{2}{(2n)!}\prod_{i=0}^{n-1}(p^2 - i^2)\right\}x^n.$$

3. Tchebysheff polynomials

The pth Tchebysheff polynomial (of the first kind) is given by

$$T_p(u) = \cos(p\cos^{-1}u).$$

Using (*) above one can show that the Adams operation ψ^p satisfies a differential equation

$$(**) \qquad (\psi^p)'' \, x\,(x+4) + (\psi^p)'(x+2) - p^2\psi^p = 2p^2.$$

The Tchebysheff polynomial T_p satisfies essentially the same equation on $[-4,0]$ and we have

THEOREM 2. $\psi^p(x) + 2 = 2T_p\left(\frac{x+2}{2}\right)$.

If we set $S_p(x) = \psi^p(x) + 2$ and use the inner product

$$\langle f,g\rangle = \int_{-4}^{0} f(x)g(x)\frac{dx}{\sqrt{-x(x+4)}}$$

then $\{S_p(x)\}$ is an orthogonal set of polynomials and $\{1, S_1(x), S_2(x), \ldots\}$ is a basis for $\mathbb{Z}[x]$ and each ψ^p is an orthogonal map.

4. A differential equation

We conclude with an indication of proof of Theorem 1. With ξ,η,y,t,x and a as above

$$c(\alpha^!\xi) = \alpha^*(c\xi) = \alpha^*(1-y) = 1-\alpha^*y = 1-t^2 = (1-t)(1+t).$$

So by definition of the Chern character ch (see [4], p. 69) we have

$$ch(\alpha^!\xi) = e^t + e^{-t} = 2 + t^2 + \frac{2}{4!}t^4 + \cdots + \frac{2}{(2n)!}t^{2n} + \cdots$$

So $ch(\xi) = 2 + y + \frac{2}{4!}y^2 + \cdots$

$$ch(x) = y + \frac{2}{4!}y^2 + \cdots + \frac{2}{(2n)!}y^n + \cdots$$

Since $f: X \to X$ has degree a we have

$$ch(f^!x) = ay + \frac{2}{4!}a^2y^2 + \cdots + \frac{2}{(2n)!}a^ny^n + \cdots$$

Now X has no torsion so ch is one-to-one and to express $f^!x$ in powers of x is the same as expressing $ch(f^!x)$ in powers of $ch(x)$. Thus we want to express

$$w(y) = ay + \frac{2}{4!}a^2y^2 + \cdots + \frac{2}{(2n)!}a^ny^n + \cdots$$

in powers of

$$z(y) = y + \frac{2}{4!}y^2 + \cdots + \frac{2}{(2n)!}y^n + \cdots .$$

We have

$$w'' = a(w+2)$$

$$(Z')^2 = Z(Z+4)$$

$$Z'' = Z+2$$

and can reduce to an equation for $w(z)$

$$(\ast\ast) \qquad w''z(z+4) + w'(z+2) - aw = 2a.$$

A particular solution is $w = -2$ and 0 is a regular singular point. The power series solution gives $(\ast)$.

REFERENCES

1. Adams-Mahmud, Maps between classifying spaces, Inventiones Math 35 (1976) pp. 1-41.

2. Atiyah-Hirzebruch, Vector bundles and homogeneous spaces, Proc. of Symp. in Pure Math. III Am. Math. Soc. (1961) pp. 7-38.

3. Bernstein, Israel, Bundles over products of infinite dimensional quaternionic projective space, Q. J. of Math, Oxford 19 (1968) 275-279.

4. Hilton, Peter, General cohomology theory and K-theory, Camb. Univ. Press (1971).

5. Hilton-Mislin-Roitberg, Localization of nilpotent groups and spaces, North Holland Mathematics Studies No. 15 (1975).

6. McGibbon, Charles, Which group structures on S^3 have a maximal torus? Preprint, 1977.

7. Milnor, John, Construction of universal bundles II, Ann. of Math 63 (1956) pp. 430-436.

8. Moser-Tangora, Cohomology operations and applications in homotopy theory, Harper and Row, 1968.

9. Rector, David, Subgroups of finite dimensional topological groups, J. Pure Appl. Alg. 1 (1971) pp. 253-273.

10. ______, Loop structures on the homotopy type of S^3, Springer Lecture Notes 249 (1971) 99-105.

11. Sullivan, Dennis, Geometric Topology, part I, MIT Notes 1970.

ON A SPACE OF GROUP ACTIONS

S. Ferry*

University of Kentucky

and the Institute for Advanced Study

A Hilbert cube manifold is a separable metric manifold modelled on the Hilbert cube. If $H(M)$ is a compact Hilbert cube manifold, we will let $H(M)$ denote the homeomorphism group of M with the topology induced from the metric $\rho(h,g) = \sup_{m \in M} d(f(m),g(m))$. If G is a finite group, an action of G on M will be regarded as a homeomorphism $\Phi\colon G \to H(M)$. The space of all actions of G on M is a metric space, $A(G,M)$, with metric $\bar{\rho}(\Phi,\Phi') = \sup_{g \in G} \rho(\Phi(g),\Phi'(g))$. We will let $FA(G,M)$ denote the space of free actions of G on M, i.e., the space of actions Φ such that $\Phi(g)\colon M \to M$ is without fixed points for each $g \neq e$.

The purpose of this note is to show that $FA(G,M)$ is a manifold modelled on the separable Hilbert space, ℓ_2. The proof of the present result is not difficult, but relies heavily on previous work of A. Edmonds, R. Geoghegan, H. Torunczyk, and the author.

If $\Phi\colon G \to H(M)$ is an action of G on M and $h \in H(M)$, let $h \cdot \Phi$ be the action such that $h \cdot \Phi(g) = h\Phi(g)h^{-1}$. We start by stating a theorem of A. Edmonds.

<u>Proposition 1</u>. If $\Phi_0 \in FA(G,M)$, then there exists a neighborhood $\mathcal{U}$ of Φ_0 in $FA(G,M)$ and a map $\psi\colon \mathcal{U} \to H(M)$ such that $\psi(\Phi) \cdot \Phi_0 = \Phi$ for each $\Phi \in \mathcal{U}$.

<u>Proof</u>: The analogous proposition for finite-dimensional manifolds is Theorem 2.1 of [E]. To prove the theorem for Hilbert cube manifolds, one need only use the main theorem of [F - V] in place of the Edwards-Kirby result on local contractibility of the homeomorphism group of a manifold. ∎

<u>Proposition 2</u>. $FA(G,M)$ is an absolute neighborhood retract.

<u>Proof</u>: Let $\mathcal{U}$, Φ_0, and ψ be as in Proposition 1. Define $c\colon H(M) \to FA(G,M)$ by $c(h) = h \cdot \Phi_0$ and let $\mathcal{V} = c^{-1}(\mathcal{U})$. The maps $p\colon \mathcal{V} \to c^{-1}(\Phi_0) \times \mathcal{U}$ and

*Partially supported by NSF Grant.

$q\colon c^{-1}(\Phi_0) \times \mathcal{U} \to \mathcal{V}$ defined by $p(h) = ([\psi(h \cdot \Phi_0)]^{-1}h,\ h \cdot \Phi_0)$ and $q(k,\Phi) = \psi(\Phi)k$ are easily seen to be inverses, so $c^{-1}(\Phi_0) \times \mathcal{U}$ is homeomorphic to $\mathcal{V}$.

$H(M)$ is known to be an ANR by [F] or [T_2]. Thus, $\mathcal{V}$ is an ANR. Since $\mathcal{U}$ is a retract of $c^{-1}(\Phi_0) \times \mathcal{U} \cong \mathcal{V}$, $\mathcal{U}$ is an ANR. ∎

Proposition 3. If $\Phi_0 \in FA(G,M)$, then there is a neighborhood $\mathcal{U}$ of Φ_0 in $FA(G,M)$ such that $\mathcal{U} \times \ell_2 \cong \mathcal{U}$.

Proof: Our proof relies heavily on ideas of Geoghegan [G]. If $f,g\colon X \times I \to Y$ are imbeddings with X compact and Y metric, we say that f is a reparameterization of g is there is a homeomorphism $k\colon (X \times I, X \times \{0\}) \to (X \times I, X \times \{0\})$ such that $k(x,t) = (x,\bar{k}(x,t))$ for some $\bar{k}\colon X \times I \to I$ and such that $f = g \circ k$. Note that if such a k exists then it is unique. A homeomorphism k as above will be called interval-preserving and the group of such homeomorphisms will be denoted by $IH(X \times I)$.

Using Marston Morse's idea of u-length, Geoghegan [G] has shown that each embedding of $X \times I$ into Y has a canonical reparameterization. Thus, if $E(X \times I, Y)$ is the space of embeddings of $X \times I$ into Y (with the compact-open topology) then there is a continuous function $R : E(X \times I, Y) \to E(X \times I, Y)$ such that $R(f)$ is a reparameterization of f and such that f is a reparameterization of g if and only if $R(f) = R(g)$.

If $\Phi_0 \in FA(G,M)$, let $i\colon Q \to M$ be a z-imbedding such that $\Phi_0(g)(i(Q)) \cap i(Q) = \emptyset$ for all $g \in G - \{e\}$. Let $F\colon Q \times I \to M$ be a closed collar on i. We can choose F so that $\mathcal{U} = \{\Phi \mid \Phi(g)(F(Q \times I)) \cap F(Q \times I) = \emptyset$ for all $g \in G - \{e\}\}$ is a neighborhood of Φ_0 in $FA(G,M)$.

Choose $g \in G - \{e\}$. g will remain fixed throughout the remainder of the proof. We will identify $F(Q \times I)$ with $Q \times I$. We will also identify $k \in IH(Q \times I)$ with the homeomorphism from M to itself which is equal to k on $Q \times I$ and which is the identity outside of $Q \times I$. If

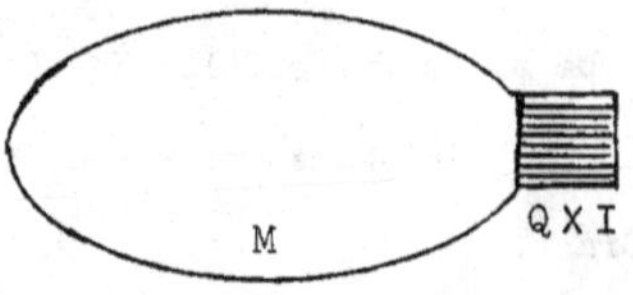

$h: M \to M$ is a homeomorphism, let $\bar{h}: M \to M$ be the homeomorphism defined by

$$\bar{h}(x) = \begin{cases} R(h \mid Q \times I)(x) & x \in Q \times I \\ h(x) & x \in M - Q \times I \end{cases}$$

$\bar{h}$ is the result of making h canonical on $Q \times I$.

We will now show how to associate a canonical action $\bar{\Phi}$ with each action $\Phi \in \mathcal{U}$. The idea is to make $\bar{\Phi}(g)$ canonical on $Q \times I$, where g is the fixed element of G chosen above. This canonical action is $\bar{\Phi} = [(\overline{\Phi(g)})^{-1}\Phi(g)] \cdot \Phi$. Note that $k = (\overline{\Phi(g)})^{-1}\Phi(g)$ is the unique element of $IH(Q \times I)$ such that $[k \cdot \Phi](g) \mid Q \times I = R(\Phi(g) \mid Q \times I)$. Thus, $\bar{\bar{\Phi}} = \bar{\Phi}$. Let $K = \{\bar{\Phi} \mid \Phi \in \mathcal{U}\}$. We have maps $p: \mathcal{U} \to K \times IH(Q \times I)$ and $q: K \times IH(Q \times I) \to \mathcal{U}$ given by $p(\Phi) = (\bar{\Phi}, \Phi(g)^{-1}\overline{\Phi(g)})$ and $q(\Phi,k) = k \cdot \Phi$. One easily checks that these two maps are inverses. Therefore, $\mathcal{U} \cong K \times IH(Q \times I)$.

Geoghehan [G] has shown that $IH(Q \times I) \cong IH(Q \times I) \times \ell_2$, so $\mathcal{U} \cong \mathcal{U} \times \ell_2$. ∎

<u>Theorem 4</u>. $FA(G,M)$ is an ℓ_2-manifold.

<u>Proof</u>: $A(G,M)$ is a closed subspace of $H(M)$ and $FA(G,M)$ is an open subspace of $A(G,M)$. Thus, $FA(G,M)$ is a separable, topologically complete ANR which admits an ℓ_2-factor locally. The main theorem of [T_1] shows that $FA(G,M)$ is an ℓ_2-manifold. ∎

The results of this paper leave several questions unanswered. We state two.

<u>Question 1</u>. Is the main result of this paper true for more general spaces of actions?

<u>Question 2</u>. Can one give reasonable conditions on a map $\alpha: M \to M$ which imply that α is close to an involution? Such conditions are known for maps which are close to homeomorphisms [F] or retractions [Ch].

REFERENCES

[B] G. Bredon, Introduction to compact transformation groups, Academic Press, New York, 1972.

[Ch] T. Chapman, The space of retractions of a compact Hilbert cube manifold is an ANR, to appear.

[E] A. Edmonds, Local connectedness of spaces of finite group actions on manifolds, Quart. Jour. Math. Oxford, 27, (1976), pp. 71-84.

[F-V] A. Fathi and Y. Visetti, Deformation of open embeddings of Q-manifolds, Trans. A.M.S. 224 (1976), pp. 427-436.

[F] S. Ferry, The homeomorphism group of a compact Hilbert cube manifold is an ANR, Annals of Math. 106, (1977), pp. 101-119.

[G] R. Geoghegan, On spaces of homeomorphisms, embeddings, and functions-I, Topology 11, (1972), pp. 159-177.

[T_1] H. Torunczyk, Absolute retracts as factors of normed linear spaces, Fund. Math., (1974), pp. 51-67.

[T_2] ______, to appear.

REMARKS ON THE SOLUTION OF FIRST DEGREE EQUATIONS IN GROUPS

Michael H. Freedman

University of California, San Diego

In the late 1950's M. Kervaire asked (see [MKS] page 403) whether a group $G \neq \{e\}$ could ever be made trivial by the addition of a single generator and relation, i.e. can $G * Z/_r = \{e\}$. Considering the ranks of the abelianized groups, $Ab(G)$ and $Ab(G * Z/_r) = Ab(G) \oplus Z/_r$ one sees that any example of G and r ($r \in G * Z$) with $G * Z/_r = \{e\}$ must satisfy 1) $G = [G,G]$ i.e., G is perfect and 2) $\deg_Z(r) = \pm 1$, where $\deg_Z(r)$ is the summation of the exponents of Z occuring in r. It follows quickly from [GR] that any such example satisfies: 3) G contains no proper normal subgroups of finite index. We will always assume that $G \neq \{e\}$ and that $\deg_Z(r) = \pm 1$.

Let z be regarded as a variable and the other letters occuring in r as constants. Suppose that G can be injected via $\beta : G \to G'$ into a group G' in which the equation $(\beta(r))(z) = 0$ has a solution s. Then if $\langle G,s \rangle$ is the subgroup of G' generated by G and s, $G \to \langle G,s \rangle$ is an injection and we have a commutative diagram with the straight compositions exact:

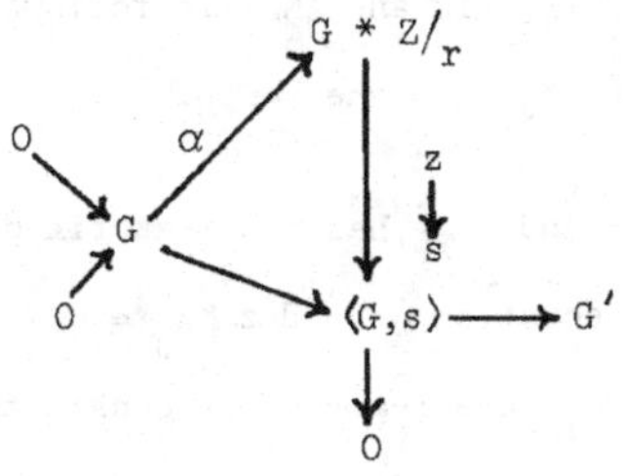

In particular, α is an injection. Similarly, if $\beta : G \to G'$ is only presumed to be non-zero (rather than an injection) and if $(\beta(r))(z)$ has a solution then α is non-zero. So Kervaire's problem may be approached as a problem of extending a group G (or merely mapping it nontrivially) to a larger group G' in which a given first degree equation $(\beta(r))(z) = 0$ has a solution. For us the most important conclusion of [GR] amounts to the discovery that any finite group G may be extended to $SO(n)$ for some n (regarding G as a set of permutation matrices) and that in $SO(n)$ any non-zero degree equation can be solved. Although Kervaire's problem has been my motivation for studying the solution of equations in groups, this study began with theorems of Hopf's on the divisibility of compact Lie groups and is of interest in its own right. See [B] and [H].

This paper is divided into five remarks.

Remark 1: Solution of first degree equations in groups of bounded isometries.

Remark 2: A first degree equation that cannot always be solved in a connected non-compact Lie group.

Remark 3: Extending groups of isometries; there is no non-compact Palais-Mostow theorem.

Remark 4: Finitely generated versus arbitrary groups G.

Remark 5: A knot theoretic motivation for Kervaire's problem.

Remark 1: Let X be a metric space and G be a path connected subgroup of the group of bounded isometries of X. So if $g \in G$, $\sup_{x \in X} d(x,g(x)) < \infty$. Let G be given the metric $d(g_1,g_2) = \sup_{x \in X} d(g_1(x), g_2(x))$ and let B_s denote the closed ball of radius s about $e \in G$. We say that B_s has the weak fixed point property (wfpp) if every map $B_s \to B_s$ which is homotopic to the constant map to e has a fixed point. For example if B_s is an ANR it follows from the Lefschetz fixed point theorem (see [B]) that B_s has the wfpp.

Theorem: If G is as above and B_s has the weak fixed point property for sufficiently large s, then any equation $r = \Pi z^{e_j} g_j = e$ can be solved for z provided $\deg_z(r) = \pm 1$. The g_i are regarded as constants in G.

Proof: The equation, that we are examining for solutions, may be rewritten

$$z = \Pi z^{e_j} g_j$$

with the $\deg_z(z^{e_j} g_j) = \Sigma e_j = 0$. Set

$$f(z) = \Pi z^{e_j} g_j$$

and

$$f_t(z) = \Pi z^{e_j} g_j$$

where $g_{jt} = \lambda(t)$ for some path $\lambda : [0,1] \to g$ from g_j to e.

$$f_0 = f \text{ and } f_1(z) = e \text{ for all } z \in G.$$

Let $d_j = \sup_t d(e, g_{jt})$. By induction on the number of occurrences of z and z^{-1} in $\Pi z^{e_j} g_j$ one may show that for all z and t, $d(f_t(z), e) \geq \Sigma d_j = s$. Applying our hypothesis to $f|B_s$ we obtain a fixed point for f and hence a solution to our equation. ||

This argument is an adaptation of arguments of Robert Brown [B] and Andrew Casson [C] for the solution of equations in compact Groups.

Any Lie group may be given a left (or right) invariant Riemannian metric and so may act on itself by left (or right) multiplication as a group of isometries. If a Lie group admits a bi-invariant Riemannian metric, then multiplication (say on the left) by g is an isometry (by left invariance), and furthermore if $\lambda(t)$ is a minimal geodesic from h to gh, then $d(h,gh) = \int_{\lambda(t)} \|\lambda'\| dt = \int_{\lambda(t)h^{-1}} \|(\lambda(t)h^{-1})'\| dt = d(e,g)$ (the middle equality by right invariance). Thus, the action of g is bounded by $d(e,g)$. So connected Lie groups with bi-invariant metrics (which include the connected compact Lie groups) satisfy the hypotheses of the theorem.

Remark 2. It is well known that the exponential map is not necessarily onto for connected non-compact Lie groups. This forces the existence of elements g for which the equations $x^n = g$, $n > 1$ have no solution. Here we present an example of a first degree equation in the group $B = \{2 \times 2$ real matrices $|a_{ij}|$ satisfying $a_{21} = 0$, $a_{11} > 0$ and $a_{22} > 0\}$ for which there is no solution in B.

Example: $z^{-2} \begin{vmatrix} 2 & 1 \\ 0 & 1 \end{vmatrix} z \begin{vmatrix} 1 & 0 \\ 0 & 2 \end{vmatrix} = \begin{vmatrix} 1 & 0 \\ 0 & 1 \end{vmatrix}$. Suppose $\begin{vmatrix} w & x \\ 0 & y \end{vmatrix}$ were a solution. Then

$\begin{vmatrix} w^2 & x(w+y) \\ 0 & y^2 \end{vmatrix} = \begin{vmatrix} 2w & 4x+2y \\ 0 & 2y \end{vmatrix}$ so $w = y = 2$ yielding $4x = 4x + 4$, a contradiction.

It follows from Remark 1 that B cannot occur as a subgroup of any G satisfying the hypotheses of the previous theorem.

Remark 3: I do not know if all finitely generated groups can be found as subgroups of groups G satisfying the hypotheses of the theorem in Remark 1. One approach is to start with a group of isometries of a metric space and try to enlarge the metric space so that the action extends to become a subgroup of the identity component of the bounded isometries of the enlarged space. For example the dihedral group, D_n, acts on the circle. If the circle is included as the equator of S^2, the action extends and is in the identity component. For another example, let Z_2 act on R by $x \mapsto -x$. This extends to an action $(x,s) \mapsto (-x,s)$ on $(R \times [0,1]/_{(x,0)=(-x,1)}$, quotient metric) which is now bounded and in the identity component. The last example can be generalized as follows:

Theorem: If A is an abelian group of rank k with generators $h_1,\ldots,h_k$ and if $H \longrightarrow \mathrm{Iso}(X)$ is a monomorphism to the isometries of a metric space, there is a monomorphism $H \longrightarrow (\text{Bounded Isometries})_{\text{identity component}}(Y)$, where $Y = ((X \times [0,1]/_{(x,0) = (h_1(x),1)}) \times [0,1]/_{(x,s_1,0) = (h_2(x),s_1,1)} \times \cdots \times [0,1]/_{(x,s_1,\ldots,s_{k-1},0) = (h_k(x),s_1,\ldots,s_{k-1},1)})$ Observe that H must be abelian for the identifications to be compatable.

Conceivably there is a construction which would work for an arbitrary H. However if X and Y above are required to be smooth manifolds we will see that there can be no such construction for an arbitrary H.

Any finitely generated group H will act freely on the universal cover of a manifold M with $\pi_1(M) = H$. Suppose H injects into the identity component of the bounded isometries of a Riemannian manifold. H then is isomorphic to a finitely generated subgroup of a Lie group G [MS]. The adjoint action gives exact sequences:

$$\begin{array}{ccccccc} 0 \longrightarrow & \text{center}(G) & \longrightarrow & G & \xrightarrow{\text{Adj}} & \mathrm{Aut}(\mathcal{L}) \\ & \uparrow & & \uparrow & & \\ 0 \longrightarrow & \text{center}(G) \cap H & \longrightarrow & H & \longrightarrow & \mathrm{Adj}(H) \end{array}$$

Adj(H) is a finitely generated subgroup of a matrix group over R and as such is residually finite (see [GR]). So if H is not a central extension of a residually finite group we cannot hope to extend the action to a manifold Y_n as in the previous theorem.

As a special case, notice that one cannot hope to find equivariant imbedding of a manifold with a finitely generated group acting (X,H) into $(R^n, GL(n,R))$ unless H is residually finite. Compare this to the Palais-Mostow theorem [P] which says that such equivariant imbeddings exist when H is finite.

<u>Remark 4</u>: Suppose we assume:

<u>Conjecture 1</u>: If H is finitely generated and r is a word in $H * Z$ with $\deg_z(r) \neq 0$ then $H \to H * Z/_r$ is an injection.

Let G be any group and r a word in $G * Z$ with $\deg_z(r) \neq 0$. Let $g_1, \ldots, g_n$ be the elements of G occurring in r and let $H = \langle g_1, \ldots, g_n \rangle \subset G$. Consider the push-out diagram:

$$\begin{array}{ccc} H & \xrightarrow{\text{inj.}} & H * Z/_r \\ \updownarrow & & \downarrow \\ G & \longrightarrow & G *_H (H * Z/_r) \end{array}$$

By conjecture 1 the top arrow is an injection, as a result the pushout is a free product with amalgamation and all four arrows are injections. If a generating set for G $(g_1, \ldots, g_n, g_{n+1}, \ldots)$ is chosen to extend our generating set for H then $G *_H (H * Z/r) = \langle g_1, \ldots, g'_1, \ldots, g'_n, z;$ relations in G, relations in H, $g_1 = g'_1, \ldots, g_n = g_n', r\rangle = \langle z, g_1, \ldots;$ relations in $G, r\rangle = G * Z_{/r}$. Since the lower arrow of the above diagram is an injection, Conjecture 1 implies:

<u>Conjecture 2</u>: If G is any group and r is a word in $G * Z$ with $\deg_z(r) \neq 0$ then $G \to G * Z/_r$ is an injection.

<u>Remark 5</u>: By a knot, k, I mean a smooth imbedding $S^1 \to S^3$. Let M_k be the 3-manifold which results from cutting out a tubular neighborhood of k and gluing it back with the meridian and longitude interchanged (0-framed surgery). If k is

the trivial knot $M_k = S^1 \times S^2$, for an arbitrary knot there is a natural map $f : M_k \to S^1 \times S^2$ (defined up to homotopy) which induces isomorphism on $H_*(\ ; Z)$. In some sense the closer k is to being trivial the closer f is to being homotopic to a homeomorphism. More precisely:

Fact 1: The Alexander polynomial $\Delta_k(t)$ is trivial $(= 1)$ iff f induces an isomorphism on

Fact 2: $H_2(\pi_1(S^3 - k)/_{\eta\langle \text{longitude} \rangle}; Z) = 0$ iff $M_k = S^1 \times S^2 \# \Sigma^3$ where Σ^3 is a homology 3-sphere.

Fact 3: If $M_k = S^1 \times S^2 \# \Sigma^3$, $\pi_1(\Sigma^3) * Z/_r$ is trivial for some relation $r \in \pi_1(\Sigma^3) * Z$.

So if a nontrivial knot k with $H_2(\pi_1(S^3 - k)/_{\eta\langle \text{longitude} \rangle}; Z)$ be found, we have found one of three interesting objects:

1) a nontrivial knot k with property R, i.e., with $M_k = S^1 \times S^2$
2) a counter example to the Poincare' conjecture that is $M_k = S^1 \times S^2 \#$ (homotopy sphere)
3) a counter example to conjecture 1 (with $H = \pi_1(\Sigma^3)$).

Fact 1 follows from a theorem of Crowell's [Cr] that says when $\Delta(t) = 1$ the infinite cyclic cover of the knot complement is acyclic. Fact 2 follows from the Shapiro-Whitehead version of the sphere theorem [SW] together with the exact sequence $\pi_2(M_k) \xrightarrow{\text{Hur.}} H_2(M_k ; Z) \longrightarrow H_2(\pi_1(M_k) ; Z) \longrightarrow 0$. Fact 3 comes from considering the trace of 0-framed surgery on k as a simply connected handle body that becomes $M_k \times I \cup$ 2-handle when turned upside down.

My impression is that Facts 2 and 3 have been independently discovered many times.

References

[B] Robert Brown, The Lefschetz Fixed Theorem, Foresman 1971.

[C] Andrew Casson, private communication.

[Cr] R. H. Crowell, The group G'/G'' of a knot group G, Duke Math. J. vol. 30 (1963), 349-354.

[GR] M. Gersteinhaber, O. S. Rothaus, The solution of sets of equations in groups, Proc. Nat. Acad. Sci. U. S. vol. 48 (1962), 1531-1533.

[H] Hopf, Uber den rang geschlossener Liescher Gruppen, Comm. Math. Helv. vol. 13 (1940-1941), 119-143.

[MKS] W. Magnus, A. Karrass, D. Solitar, Combinatorial Group Theory, New York Interscience Publishers, 1966.

[MS] S. B. Myers, N. E. Steenrod, The group of isometries of a Riemannian manifold, Ann. of Math. vol. 40 (1939), 400-416.

[P] R. Palais, Embedding compact differentiable transformation groups in orthogonal representations, J. Math. Mech. 6 (1957), 673-678.

[SW] Shapiro, Whitehead, A proof and extension of Dehn's lemma, Bull. A.M.S. vol. 64 (1958), 174-179.

FLAT MANIFOLDS AND THE COHOMOLOGY OF GROUPS

Morris W. Hirsch

University of California, Berkeley

§1. Introduction

Let M be a complete flat manifold. This means there is a specified covering space $f: M \to R^n$ whose group π of deck transformations operates on R^n by affine automorphisms. (Incomplete flat manifolds are discussed in Section 4).

Let $U \subseteq M$ be an open set and $\Phi: U \to R^n$ be a coordinate chart. (Φ, U) is called affine if $\Phi \circ f^{-1}: f(U) \to R^n$ is the restriction of an affine map. Clearly M is covered by such charts.

Let U be as above and let X, Y be vectors tangent to U, perhaps at different points. When expressed in affine coordinates, they define vectors $(X_1, \dots X_n)$, $(Y_1, \dots, Y_n) \in R^n$. If $X_i = Y_i$, $i = 1, \dots, n$ we say that X and Y are parallel vectors. This property is independent of the choice of affine coordinates.

A vector field X on M is parallel if X_p and Y_p are parallel whenever p and q lie in the domain of an affine coordinate system. Equivalently, X is f-related to a constant vector field on R^n.

Let $H: \pi \hookrightarrow \text{Aff}(n)$ be the inclusion of π in the group of affine automorphisms of R^n. Let $\mathscr{L}: \text{Aff}(n) \to GL(n)$ be the homomorphism that assigns to each affine automorphism its linear part. Then

$$\mathscr{L} \circ H = h: \quad \pi \to GL(n)$$

is a homomorphism, called the (linear) holonomy representation of π. The group $h(\pi)$ is called the holonomy group of M.

There is a natural isomorphism between the vector subspace of R^n consisting of stationary points of the action of $\mathscr{L}(\pi)$, and the vector space of all parallel vector fields on M.

THEOREM A. Let M be a complete flat manifold with nilpotent fundamental group.

Then M admits a nonzero parallel vector field.

This will be proved in Section 2 with the aid of a result proved in Section 3 on the cohomology of nilpotent groups.

Besides vector fields, other kinds of parallel geometric objects on M can be defined. For example a field F of tangent k-planes is called __parallel__ if it is f-related to a constant field of k-planes on R^n. Such a field F is integrable, and defines a __parallel__ foliation of M.

Let $U \subseteq M$ be the domain of an affine coordinate system $(x_1,\ldots,x_n)$. If $\mathfrak{F}$ is a foliation of codimension one then there is a 1-form $\omega_U = \Sigma a_i\, dx_i$ defined on U such that in U, $\mathfrak{F}$ is determined by $\omega_U = 0$. If $\mathfrak{F}$ is parallel, then the ω_U can be chosen with the a_i constant.

It may happen that there is a cover $\mathcal{U}$ of M by affine coordinate domains and 1-forms ω_U, $U \in \mathcal{U}$ as above such that $\omega_U = \omega_V$ on $U \cap V$ for all $U,V \in \mathcal{U}$. In this case there is a global 1-form ω on M such that $\omega_U = \omega$, and $\mathfrak{F}$ is defined by $\omega = 0$. Such an $\mathfrak{F}$ is called a __measured parallel foliation.__ Since each ω_U is evidently closed, ω is closed. Moreover ω lifts to a 1-form $\tilde{\omega}$ on R^n with constant coefficients invariant under the natural action of π on 1-forms. Such a 1-form $\tilde{\omega}$ is the same as a linear map $L\colon R^n \to R$ invariant under the adjoint action of the holonomy group.

THEOREM B. Let $M = R^n/\pi$ be a complete flat manifold with nilpotent fundamental group π. Suppose the holonomy group $h(\pi)$ preserves a nondegenerate bilinear form Ω on R^n; then M admits a measured parallel foliation.

PROOF. By Theorem A, there exists a nonzero vector $X \in R^n$ which is stationary under $h(\pi)$. The linear map $L\colon R^n \to R$, $L(Y) = \Omega(X,Y)$ defines a 1-form on R^n invariant under the adjoint action of $h(\pi)$. Therefore L corresponds to a measured parallel foliation on M. It is well-known that every compact manifold which admits a nowhere vanishing closed 1-form ω fibers over S^1. (Let θ be a closed 1-form near ω, having rational periods: Some nonzero integer multiple $m\theta$ has integer periods. Integrating the lift of $m\theta$ over the paths from $0 \in R^n$

defines a map $R^n \to R$ which covers a fibration $R^n/\pi \to R/Z$.) Q.E.D.

EXAMPLE. The Klein bottle K has a complete affine structure whose holonomy group is generated by the affine map $(x,y) \mapsto (x+1, -y)$. The foliation of R^2 by the lines $x =$ constant is the lift of a measured parallel foliation of K; the corresponding constant 1-form $L: R^n \to R$ is $(x,y) \to x$. The foliation of R^2 by the lines $y =$ constant is the lift of a parallel foliation which is not measured since it is not transversely orientable.

Many explicit constructions of 3-dimensional flat manifolds satisfying the hypotheses of Theorems A and B are found in Auslander [1], Auslander and Markus [2].

In the case of a compact flat <u>Riemannian</u> manifold (i.e. π acts on R^n by isometries), it is well known that the dimension of the space of parallel vector fields equals the first Betti number of M. This result was stated parenthetically by Calabi [3].

§2. Cohomology of groups and the proof of Theorem A.

Let M be as in Theorem A; denote its fundamental group by π. Let E denote the π-module whose underlying group is R^n, the action of π being defined by the holonomy representation. If π is the trivial group. Theorem A is obvious. Henceforth assume π is nontrivial.

LEMMA 1. $H^1(\pi,E) \neq 0$.

PROOF: Every $\gamma \in \pi$ is an affine automorphism of R^n whose linear part is $h(\gamma) \in GL(n)$. Let $u(\gamma) \in R^n$ be the translational part of γ. Then

$$\gamma x = h(\gamma)x + u(\gamma) \tag{1}$$

for all $\gamma \in \pi$, $x \in R^n$. It is easy to see that

$$u(\gamma\delta) = h(\gamma)\, u(\delta) + u(\gamma) \tag{2}$$

Thus u is a crossed homomorphism for h, and so u defines an element c_M of $H^1(\pi,E)$. If $c_M = 0$ then u has the form

$$u(\gamma) = y - h(y) \qquad (\gamma \in \pi)$$

for some $y \in R^n$. Then (1) would imply $\gamma y = y$ for all $\gamma \in \pi$. But every nontrivial element of π operates without fixed points. Therefore $c_M \neq 0$. Q.E.D.

In Section 3 I will prove:

THEOREM C. Let G be a nilpotent group acting linearly on a finite dimensional vector space; denote the resulting G-module by E. If $H^0(G,E) = 0$, then $H^i(G,E) = 0$ for all $i \geq 0$.

It follows from the Lemma and Theorem C that $H^0(\pi,E) \neq 0$. By definition, $H^0(\pi,E)$ is the set of vectors of R^n stationary under $h(\pi)$. Therefore M has a parallel vector field. This proves Theorem A.

§3. Proof of Theorem C.

Let G be a group. By a G-module M we mean a vector space M (over any field) together with a linear action of G on M.

The theory of the cohomology groups $H^i(G,M)$ is expounded in Atiyah-Tate [7] or MacLane [8]. For present purposes the following definitions suffice. The subspace of vectors in M fixed under every element of G is denoted by both M^G and $H^0(G,M)$. Define $H^1(G,M)$ to be the quotient vector space Z^1/B^1 where Z^1 is the space of crossed homomorphisms

$$u: G \to M, \qquad u(\gamma\delta) = \gamma u(\delta) + u(\gamma)$$

and B^1 consists of those u having the form, for some $y \in M$,

$$u(\gamma) = (1 - \gamma)y = y - \gamma y$$

The higher cohomology groups can be defined recursively as follows. Define M' to be the vector space of all maps $\Phi: G \to M$ with the usual pointwise operations. G acts on M' by

$$\gamma\Phi: G \to M, \qquad \delta \to \gamma(\Phi(\delta))$$

for $\gamma \in G$, $\delta \in G$, $\Phi \in M'$. An embedding $i: M \hookrightarrow M'$ of G-modules is defined by

$$i(x): G \to M, \qquad \gamma \mapsto \gamma x$$

Set $M^{\#} = M^{\#}_{G} = M'/i(M)$. Then define $H^2(G,M) = H^1(G,M^{\#})$; recursively define $H^k(G,M) = H^{k-1}(G,M^{\#})$, $k \geq 2$.

It turns out that M' is <u>acyclic</u>, i.e. $H^i(G,M') = 0$ for all $i > 0$.

LEMMA 2. Let M be any G-module. Suppose there exists γ in the center of G such that $1 - \gamma$ is invertible as a linear operator on M. Then $H^k(G,M) = 0$ for $k = 0,1$.

PROOF: $H^0(G,M) = 0$ because γ has only one fixed point. If $u: G \to M$ is a crossed homomorphism, define an affine action A of G on M by

$$A(\alpha): M \to M$$
$$x \mapsto \alpha x + u(\alpha), \qquad (\alpha \in G).$$

Then

(3) $$A(\alpha\beta) = A(\alpha) \circ A(\beta).$$

It is easy to see that $A(\gamma)$ has a unique fixed point $y \in M$; solving $A(\gamma)y = y$ one finds that $y = (1 - \gamma)^{-1}u(\gamma)$. Since γ commutes with every element of G, it follows from (3) (with $\beta = \gamma$) that y is fixed under every $A(\alpha)$, $\alpha \in G$. This means that $u(\alpha) = (1 - \alpha)y$. Hence u represents $0 \in H^1(G,M)$. Q.E.D.

LEMMA 3. Let G,M be as in Lemma 2. Then $H^k(G,M) = 0$ for all $k \geq 0$.

PROOF. By induction on k, the cases $k = 0,1$ having been proved in Lemma 2. Let $k \geq 2$ and assume that it is known that $H^i(G,M) = 0$ for all $i < k$ whenever the hypothesis of Lemma 2 holds.

Consider $H^k(G,M)$. This is the same as $H^{k-1}(G,M^{\#})$ by our definition. It is easy to verify that the operator $1 - \gamma$ on M' is invertible, and therefore $1 - \gamma$ is invertible on $M^{\#}$. The inductive hypothesis now implies that $H^{k-1}(G,M^{\#}) = 0$. Q.E.D.

LEMMA 4. Theorem C is true if G is abelian.

PROOF: Let E be a vector space over a field k, of finite dimension $n \geq 0$. Let ℓ be the algebraic closure of k; then $E \otimes \ell$ is an n-dimensional vector space ℓ; the action of G extends naturally to an action on $E \otimes \ell$, and $H^i(G, E \otimes \ell) = H^i(G,E) \otimes \ell$. Thus it suffices to prove Lemma 4 under the assumption that E is a vector space over an algebraically closed field ℓ.

The case $\dim E = 0$ being trivial, we assume $\dim E = n > 0$ and proceed by induction on n.

Suppose E is a simple G-module: no nontrivial proper linear subspace is G-invariant. Since ℓ is algebraically closed and G is abelian, it follows from Schur's lemma, that $\dim E = 1$. Since $H^0(G,E) = 0$, G does not act trivially. Since $\dim E = 1$, this means that $1 - g$ is invertible for some $g \in G$. Therefore when E is simple the conclusion of Lemma 4 follows from Lemma 3.

If E is not simple, set $F \subset E$ be a G-invariant subspace with $0 < \dim F < \dim E$. Then F and E/F are G-modules. The exact sequence $0 \to F \to E \to E/F \to 0$ yields the exact sequence

$$0 \to H^0(G,F) \to H^0(G,E) \to H^0(G,E/F) \to H^1(G,F) \to \cdots \to H^i(G,F) \to H^i(G,E)$$

$$\to H^i(G,E/F) \to H^{i+1}(G,F) \cdots$$

Since $H^0(G,E) = 0$ exactness shows $H^0(G,F) = 0$. The induction hypothesis makes $H^i(G,F) = 0$ for all i. In particular $H^1(G,F) = 0$ forces $H^0(G,E/F) = 0$ and induction now makes $H^i(G,E/F) = 0$ for all i. Therefore $H^i(G,E) = 0$ for all i. This completes the inductive proof of Lemma 4.

One way of defining a group G to be nilpotent is as follows: there exists an integer $\nu = \nu(G) \geq 0$, called the <u>nilpotence</u> of G, and normal subgroups $G_i \subset G$, $0 \leq i \leq \nu$, forming the <u>lower central series</u>

$$G = G_0 \supset G_1 \supset G_2 \supset \cdots \supset G_\nu,$$

such that $G_\nu = C(G) =$ the center of G, and $G_{i-1}/G_i = C(G/G_i)$ for $1 \leq i \leq \nu$.

Thus $\nu = 0$ means that G is abelian; $\nu = 1$ means $G/C(G)$ is abelian. Notice that $G_1[G,G]$ = the commutator subgroup, and $G_{i+1} = [G,G_i]$.

LEMMA 5. Let G be nilpotent and nonabelian. Let $g \in G$ be any element and let Γ be the subgroup generated by g and G_1. Then Γ is normal and $\nu(\Gamma) < \nu(G)$.

PROOF: By induction on ν. If $\nu = 1$, G_1 is the center of G and G/G_1 is abelian. If $a \in G$ then $aga^{-1} \equiv g \bmod G_1$; this implies that Γ is normal. Clearly Γ is abelian so $\nu(\Gamma) = 0 < \nu(G)$. For the inductive step assume $\nu \geq 2$ and let $\bar{\Gamma} \subset G/G_\nu$ be the subgroup generated by G_1/G_ν and gG_ν. The inductive hypothesis is: $\bar{\Gamma}$ is normal and $\nu(\bar{\Gamma}) < \nu(G/G_\nu) - 1$. Now Γ is the inverse image of $\bar{\Gamma}$ under the canonical homomorphism $G \to G/G_\nu$; therefore Γ is normal. Also $\nu(G/G_\nu) = \nu(G) - 1$, and it is easy to see that $\nu(\Gamma) \leq \nu(\bar{\Gamma}) + 1$. Thus $\nu(\Gamma) < \nu(G) - 1$. Q.E.D.

PROOF OF THEOREM C. The theorem is trivial if $\dim E = 0$ and has been proved in Lemma 4 for the abelian case $\nu(G) = 0$. We assume $\dim E = n > 0$ and $\nu(G) = \nu > 0$ and proceed by induction on $\min(n,\nu)$.

If E is not a simple G-module then the induction hypothesis on ν, and the same exactness argument used in the proof of Lemma 4, shows $H^1(G,E) = 0$ for all i.

Suppose finally that E is simple. Since $\dim E > 0$ and $H^0(G,E) = 0$, there exists $g \in G$ acting nontrivially. By Lemma 5, g belongs to a normal subgroup $\Gamma \subset G$ with $\nu(\Gamma) < \nu(G)$. Normality of Γ implies that E^Γ is G-invariant. Hence $E^\Gamma = 0$ or E by simplicity. But g, hence also Γ, operates nontrivially so $H^0(\Gamma,E) = 0$. The inductive hypothesis on ν yields $H^j(\Gamma,E) = 0$ for all j.

The spectral sequence of Lyndon (see e.g. MacLane [8]) has

$$E_2^{i,j} \cong H^i(G/\Gamma,\ H^j(\Gamma,E)) \underset{j}{\Rightarrow} H^{i+j}(G,E),$$

converging as shown to the cohomology of G. It follows that $E^2_{i,j} = 0$ for all i,j, proving that $H^k(G,E) = 0$ for all k. The proof of Theorem C is complete. Q.E.D.

REMARK. All that is needed for the proof of Theorem B is that $H^1(G,E) = 0$ which can be proved without spectral sequences, using instead the exact "inflation-restriction" sequence

$$0 \to H^1(G/\Gamma, E^\Gamma) \to H^1(G,E) \to H(\Gamma,E).$$

§4. Incomplete flat manifolds.

A (not necessarily complete) flat manifold is a manifold M together with an atlas of coordinate charts all of whose coordinate changes extend to affine automorphisms of R^n. The definition of parallel vector fields goes through as before.

Let $\tilde{M} \to M$ be a universal cover. Let π be the group of deck transformations. There is defined an immersion $F: \tilde{M} \to R^n$ (where $n = \dim M$) and a homomorphism $H: \pi \to \mathrm{Aff}(n)$ into the affine group on R^n, such that F is equivariant respecting H; and F, H are unique up to affine conjugacy.

Let $u: \pi \to R^n$, $h: \pi \to GL(n)$ be defined by

$$H(\gamma)x = u(\gamma) + h(\gamma)x$$

Then h is a homomorphism and u is a crossed homomorphism for the π-module structure on R^n defined by h. Let

$$c_M \in H^1(\pi, R^n)$$

denote the cohomology class containing u.

Suppose that π is nilpotent. In Lemma 1 it was proved that $c_M \neq 0$ if M is complete; all that was really needed for the proof was that $H^1(\pi, R^n) \neq 0$. The rest of Theorem A goes through under this assumption:

THEOREM D. Let M be an incomplete n-dimensional flat manifold with nilpotent fundamental group. If $H^1(\pi, R^n) \neq 0$ for the holonomy representation of π on R^n, then M has a nonzero parallel vector field.

A theorem in the converse direction is

THEOREM E. Let π be the fundamental group of an n-dimensional flat manifold M. Suppose that there is an epimorphism $\pi \to Z$. If $H^1(\pi, R^n) = 0$ then M does not have a nonzero parallel vector field.

PROOF: Let E denote the π-module structure on R^n defined by $h: \pi \to GL(n)$. By assumption $H^1(\pi,E) = 0$, and we are to prove $H^0(\pi,E) = 0$. Let $0 \to K \to \pi \to Z \to 0$ be exact. There is a natural linear action of Z on E^K (the stationary set of K on E) and $H^0(\pi,E) = H^0(Z,E^K)$. From the exactness of $0 \to H^1(Z,E^K) \to H^1(\pi,E)$ we have $H^1(Z,E^K) = 0$. If $1 \in Z$ acts on E^K by $A \in GL(E^K)$ then this implies $I - A$ is invertible. Therefore $H^0(Z,E^K) = 0$. Q.E.D.

§5. Solvable holonomy.

One can get results somewhat weaker than Theorem A when the fundamental group, or more generally the holonomy group, is solvable.

THEOREM F. Let M be an n-dimensional flat manifold whose holonomy group Γ is solvable. Then M has a field of parallel j-planes $\mathfrak{F}_j$ and a field of parallel k-planes $\mathfrak{F}_k$, where $j \in \{1,2\}$ and $k \in \{n-1, n-2\}$.

PROOF: Complexify the action of Γ. Since Γ is solvable there is a 1-dimensional subspace $E \subset C^n$ invariant under Γ. If $x,y \in R^n$ are such that $x + iy$ spans E than the foliation $\mathfrak{F}_j$ corresponds to the subspace of R^n spanned by x and y. And there is a 1-dimensional subspace E' of $\mathrm{Hom}_C(C^n,C)$ invariant under the adjoint action of Γ. If $\theta,\omega: R^n \to R$ are linear maps such that $\theta + i\omega$ spans E', then the subspace $\ker\theta \cap \ker\omega \subset R^n$ corresponds to a parallel foliation $\mathfrak{F}_k$. Q.E.D.

REFERENCES

1. Auslander, L. Examples of locally affine spaces, Ann. of Math. vol 64 (1956) pp. 255-259.

2. Auslander, L and Markus, L. Flat Lorentz 3-manifolds, Amer. Math. Soc. Memoir 30, (1959).

3. Calabi, E. Closed, locally euclidean 4-dimensional manifolds, Bull. Amer. Math. Soc. vol. 63 (1975) p. 135.

4. Hirsch, M. and Thurston, W. Foliated bundles, flat manifolds and invariant measures, Ann. of Math. 101 (1975) pp. 369-390.

5. Milnor, J. On the fundamental groups of complete affinely flat manifolds, Inst. Adv. Study (Princeton), preprint.

6. Kostant, B. and Sullivan, D. The Euler characteristics of an affine space form is zero, Bull. Amer. Math. Soc. 81 (1975), p. 937.

7. Atiyah, M. and Tate, J. In Cassels and Fröhlich (Eds.) Algebra Number Theory, Thompson Book Company, Washington, 1967.

8. Mac Lane, Homology Theory, Springer-Verlag, Berlin, 1963.

TWO CHARACTERISTIC CLASSES AND SMITH THEORY

by Lowell Jones

State University of New York at Stony Brook

Introduction

Notation. p = positive odd prime, Z_p = additive group of order p, Z_p = multiplicative group of order p, $Z_{(2)}$ = integers localized at 2, $Z_{(p)}$ = integers localized at p, Q = rational numbers, $Q(Z_p)$ = rational group ring, $W(Q)$ or $W(Q(Z_p))$ = Witt-Grothendiek group of nonsingular symmetric forms or non-singular hermitian forms on f.g. Q-modules or $Q(Z_p)$-modules respectively (for the definition of hermitian form, see 2.0 below).

Let $\varphi: Z_p \times M \to M$ denote a PL group action defined on the manifold pair $M, \partial M$. $K, \partial K$ denotes the fixed point set of φ. By a theorem of P.A. Smith [16] $K, \partial K$ must be a Z_p-homology manifold pair. In the rest of this paper, it will be assumed that M and K come equipped with integral-homology orientation classes $[M], [K]$ and that the codimension of K in M equals $0 \bmod 4$.

$\varphi: Z_p \times R, R_\partial \to R, R_\partial$ denotes an equivariant neighborhood for $K, \partial K$ in $\varphi: Z_p \times M, \partial M \to M, \partial M$. In §1 below, two characteristic classes are constructed:

$$\sum_i \gamma^i(R) \in \sum_i H^{4i}(R/\dot{R}, W(Q) \otimes_Z Z_{(2)})$$

and

$$\sum_i \sigma^i(\varphi,R) \in \sum_i H^{4i}(R/\dot{R}, W(Q(Z_p)) \otimes_Z Z_{(2)}),$$

where $\dot{R}$ is the topological boundary of R in M. $\gamma^*(R)$ depends only on the PL topological type of the neighborhood R, but $\sigma^*(\varphi,R)$ depends as well on the action $\varphi: Z_p \times R \to R$. These characteristic classes are constructed by the transversality techniques of J. Morgan and D. Sullivan [13].

The main result of this paper is that there exists in general a relationship between the classes γ^* and σ^* (see Theorem 2.4 in §2 below).

This relationship is just a characteristic class version of the PL equivariant index theorem proved in [9]. This PL equivariant index theorem is stated in §2 below.

As an application of our main result, we get

Proposition 0.1. $\exists$ a characteristic class $\sum_i h_i^p(K) \in \sum_i H_{4i-1}((K,\partial K),Z)$, defined for any integrally-oriented Z_p-homology manifold pair K, ∂K, satisfying:

a) $2 \cdot h_i^p(K) = 0 \ \forall i$.

b) $h_i^p(K) = 0 \ \forall i$ if K is the fixed point set of a PL action $\varphi: Z_p \times M \to M$ defined on an orientable PL manifold M.

c) $h_i(K) = 0 \ \forall i$ if K, ∂K is an integral homology manifold pair.

d) If $g: K_1, \partial K_1 \to K_2, \partial K_2$ is a homeomorphism (not necessarily PL), then $g^*(\sum_i h_i^p(K_1)) = \sum_i h_i^p(K_2)$.

e) $\exists$ an integrally oriented Z_p-homology manifold pair K, ∂K with $\overline{H}_*(K,Z_p) = 0$, but $\sum_i h_i^p(K) \neq 0$.

Remark. P. A. Smith has proved in [16] that if $\varphi: Z_p \times B^m \to B^m$ is a PL action on the m-dimensional ball, then the fixed point set K satisfies

(i) K, ∂K is a Z_p-homology manifold pair $(\partial K \equiv K \cap \partial B^m)$.

(ii) $\overline{H_*}(K,Z_p) = 0$.

b) and e) in Proposition 0.1 show that there are further properties that K must satisfy that cannot be deduced from Smith's results.

Credits. Most of the results in this paper represent one section of a long, unpublished paper by the author [5]. Among these number Proposition 0.1 above, which was first announced in [6], and the PL equivariant index theorem (see 2.1 below), which was first announced in [7]. The PL equivariant index theorem of §2 below has been substantially extended by J. Alexander and G. Hammerick [2]. Their work, which extends previous results with J. Vick [1], first derived its direction from a result of P. E. Conner and F. Raymond [3].

Section 1. In this section the characteristic classes γ^*, σ^* are defined. Some basic properties of γ^* are discussed. The main tool of this section is the construction of characteristic classes by transversality methods, as described in Section 7 of J. Morgan's and D. Sullivan's paper [13].

First I will construct the characteristic class $\sum_i \gamma^I(K) \in \sum_i H^{4i}(R/\dot{R}, W(Q) \otimes_Z Z_{(2)})$. Only the following assumptions shall be used.

1.1. (a) K, ∂K is an integrally oriented, rational homology manifold pair.

(b) R, R_∂ is an oriented PL manifold thickening of K, ∂K, and $\dot{R} = \overline{\partial R - R_\partial}$.

(c) The codimension of K in R is equal to zero mod 4.

Let $\Omega_*(\ )$ denote the bordism functor for oriented differentiable manifolds. For each $i > 0$ a homomorphism,

1.2. $a_i : \Omega_{4_i}(R/\dot{R}) \to W(Q)$ is constructed as follows. Represent $x \in \Omega_{4_i}(R,\dot{R})$ by $g : N \to R/\dot{R}$. Push $g(N)$ away from ∂K; then put g in transverse position to every simplex of K. Then $g^{-1}(K)$ is a finite simplicial complex which is a Q-homology manifold (because N is a closed differentiable manifold and K, ∂K is a Q-homology pair). Since K, R, N are all integrally oriented, by $[K]$, $[R]$, $[N]$ say, $g^{-1}(K)$ is also integrally oriented by a class $[g^{-1}(K)]$ determined uniquely from $[K]$, $[R]$, $[N]$. Also $\dim(g^{-1}(K)) = 0 \bmod 4$ because $\dim(R) = \dim(N) = \dim(K) = 0 \bmod 4$. Write $\dim(g^{-1}(K)) = 4\ell$, and let $a_i(X) \in W(Q)$ be the element represented by the intersection pairing

$$H_{2\ell}(g^{-1}(K),Q) \times H_{2\ell}(g^{-1}(K),Q) \to Q .$$

Similarly, homomorphisms

1.3. $a_{i,q} : \Omega_{4i}(R/\dot{R}, Z|_q) \to W(Q) \otimes_Z Z_q$ can be constructed for all $i > 0$ and $q = 2^n$, $n = 1, 2, 3, \ldots$. Here $\Omega_*(\ , Z|_q)$ denotes the bordism groups of oriented differentiable $Z|_q$ manifolds defined in Section 1 of [13]. In this case $x \in \Omega_{4i}(R/\dot{R}, Z|_q)$ is represented by $g : N \to R/\dot{R}$ where N is a $Z|_q$ manifold (see §1 of [13]). So $g^{-1}(K)$ is not a rational homology manifold, but rather is obtained from an integrally oriented rational homology manifold pair $(W,\partial W)$ be identifying q boundary components $\partial_1 W$, $\partial_2 W$, ..., $\partial q W$ of ∂W (with $\partial W = \bigcup_i \partial_i W$) by orientation preserving PL homeomorphisms $\partial_1 W \to \partial_2 W \to \dots \to \partial q W$. Let $\lambda : H_{2\ell}(W,Q) \times H_{2\ell}(W,Q) \to Q$ be the middle dimension intersection form. Set $A = \{y \in H_{2\ell}(W,Q) | \lambda(y,y') = 0 \ \forall y' \in H_{2\ell}(W,Q)\}$. Then $\lambda : \frac{H_{2\ell}(W,Q)}{A} \times \frac{H_{2\ell}(W,Q)}{A} \to Q$ is a well-defined <u>non-singular</u> bilinear form representing an element $\bar{a}_{i,q}(x) \in W(Q)$. That $\bar{a}_{i,q}(x)$ is well defined modulo elements of $W(Q)$ divisible by q is argued as in §1 of [13]. [Note: What precisely is needed here to extend the arguments of §1 in [13] to the present case is an extension of Theorem 1.2 in [13]--called Novikov's additivity theorem--to cover the cases when (in the notation of [13]) M_1^n, M_2^n are oriented PL rational homology manifolds and $I(M_i^n)$ is the element of $W(Q)$ represented by the intersection pairing

$$\lambda : \frac{H_{n/2}(M_i^n,Q)}{A_i} \times \frac{H_{n/2}(M_i^n,Q)}{A_i} \to Q$$

where $A_i \equiv \{x \in H_{n/2}(M_i^n,Q) | \lambda(x,y) = \forall y \in H_{n/2}(M_i^n,Q)\}$.] Now define $a_{i,q}(x) \in W(Q) \otimes_Z Z_q$ to equal the image of any $\bar{a}_{i,q}(x)$ under the quotient map $W(Q) \to W(Q) \otimes_Z Z_q$.

The homomorphisms of 1.2, 1.3, commute with the change of coefficient maps $Z \to Z_q$, $Z_q \to Z_q$, and satisfy $a_i(y \cdot x) = \text{index}(y) \cdot a_i(x)$ or $a_{i,q}(y \cdot x) = \text{index}(y) \cdot a_{i,q}(x)$ for $y \in \Omega_*(\text{pt.})$ and $x \in \Omega_{4i}(R/\dot{R})$ or $x \in \Omega_{4i}(R/\dot{R},Z|_q)$. So as in §7 of [13], the maps $\{a_i \otimes_Z Z_{(2)}, a_{i,q}\}$ correspond under a universal coefficient theorem to a characteristic class $\gamma^i(K) \in H^{4i}(R/R_\partial, W(Q) \otimes_Z Z_{(2)})$.

<u>The classes</u> $\gamma_T^i(K)$, $\gamma_Z^i(K)$.

$\text{ind} : W(Q) \to Z$ sends each quadratic form to its index. <u>ind</u> has a left inverse $r : Z \to W(Q)$ which sends $n > 0$ to $\underbrace{[1] \oplus [1] \oplus \dots \oplus [1]}_{n\text{-fold}}$, and sends $n < 0$ to $\underbrace{[-1] \oplus [-1] \oplus \dots \oplus [-1]}_{n\text{-fold}}$. Thus $W(Q) = W_T(Q) \oplus Z$, where $W_T(Q) = \text{kernel(ind)}$. $W_T(Q)$ is the torsion subgroup of $W(Q)$ and has exponent equal to 4 ([4]). Let

1.4. $a_i = a_{i,T} \oplus a_{i,Z}$; $a_{i,q} = a_{i,q,T} \oplus a_{i,q,z}$ be the corresponding splittings of the homomorphisms in 1.2, 1.3. The collections $\{a_{i,T}, a_{i,q,T}\}$, $\{a_{i,Z}, a_{i,q,Z}\}$ yield characteristic classes $\gamma_T^i(K) \in H^{4i}(R/\dot{R}, W_T(Q))$, $\gamma_Z^i(K) \in H^{4i}(R/\dot{R}, Z_{(2)})$, satisfying

1.5. $\gamma^i(K) = \gamma_T^i(K) \oplus \gamma_Z^i(K)$.

Properties of $\sum_i \gamma_T^i(K)$ and $\sum_i \gamma_Z^i(K)$.

$\sum_i \gamma_Z^i(R)$ is the transversality characteristic class constructed by J. Morgan and D. Sullivan in §7 of [13].

Concerning $\sum_i \gamma_T^i(R)$, we have

Proposition 1.6. $4 \cdot \gamma_T^i(R) = 0 \; \forall i$.

Proposition 1.7. *If* K, ∂K *is an integral homology manifold pair, then* $\gamma_T^i(R) = 0$, $\forall i$.

Conjecture 1.8. *The hypothesis of* 1.7 *can be weakened to assuming that* K, ∂K *is a codimension* 2 PL *subset of an integral homology manifold pair* M, ∂M.

Question 1.9. *Let* $\gamma_i^T(R) \in H_{r-4i}((K,\partial K), W_T(Q))$ *denote the class which* $\gamma_T^i(R)$ *is mapped to under the composition of Poincaré duality* $H^{4i}(R/\dot{R}, W_T(Q)) \cong H_{r-4i}(R/R_\partial, W_T(Q))$ *with the homotopy retraction map* $H_{r-4i}(R/R_\partial, W_T(Q)) \cong H_{r-4i}((K,\partial K), W_T(Q))$. *Here* r = dimension (R). *If* R_1, R_2 *are two* PL *neighborhoods for* K, *is* $\sum_i \gamma_i^T(R_1) = \sum_i \gamma_i^T(R_2)$?

Proposition 1.10. *Let* (R, R_∂), (R', R'_∂) *be the stable regular neighborhoods for* K, ∂K, K', $\partial K'$ in B^N, ∂B^N (N = *large*), *where* B^N *is the* N-*dimensional ball. Then if* $h : (K,\partial K) \to (K',\partial K')$ *is a homeomorphism (not necessarily* PL) *we have* $h_*(\sum_i \gamma_i^T(R)) = \sum_i \gamma_i^T(R')$.

Proof of 1.6. The exponent of $W_T(Q)$ is 4, so the exponent of $H^*(R/\dot{R}, W_T(Q))$ is 4.

Proof of 1.7. It suffices to show that the homomorphisms $a_{i,T}$, $a_{i,q,T}$ of 1.4 are the zero maps.

First consider $a_{i,T} = 0$. Returning to the definition of $a_i(x)$, we see that $g^{-1}(K)$ must be an integral homology manifold because K, ∂K is an integral homology manifold pair. So if $[b_{ij}]$ is a matrix representation for the intersection pairing

$$\frac{H_{2\ell}(g^{-1}(K))}{\mathrm{Tor}(H_{2\ell}(g^{-1}(K)),Z)} \times \frac{H_{2\ell}(g^{-1}(K))}{\mathrm{Tor}(H_{2\ell}(g^{-1}(K)),Z)} \to Z ,$$

then $b_{ij} \in Z$ and $\det[b_{ij}] = \pm 1$. By Theorems 1 and 2 of [11],

$$T\left([b_{ij}] \oplus \begin{bmatrix} 0 & 1 \\ 1 & 0 \end{bmatrix}\right)T^t = \begin{bmatrix} \lambda_1 & & & \bigcirc \\ & \lambda_2 & & \\ & & \ddots & \\ \bigcirc & & & \lambda_n \end{bmatrix}$$

where $\lambda_i = \pm 1$ and T is in $GL(Q,n)$ with $\det(T) \neq 0$. Such matrices $[b_{ij}]$ represent elements in $W(Q)$ which have no torsion component in the splitting $W(Q) = W_T(Q) \oplus Z$. Since $[b_{ij}]$ represents $a_i(x)$ it follows that $a_{i,T}(x) = 0$.

Now consider $a_{i,q,T} = 0$. Returning to the construction of $a_{i,q}(x)$, it would suffice to show that $(W, \partial W = \bigcup_{i=1}^{q} \partial_i W)$ can be chosen to be an integral homology Poincaré duality pair with each $\partial_i W$ equal to an integral homology sphere. For then $a_{i,q}(x)$ is represented by $[b_{ij}]$ as in the preceding paragraph. We shall not prove quite this, but we shall prove that W, ∂W can be modified in the category of integral algebraic Poincaré duality chain complexes (see [12]) to $(W', \partial W')$, so that the intersection forms of W' and W represent the same elements in $W(Q) \otimes_Z Z_q$, and $\partial W'$ is a union of integral homology spheres. To get W', $\partial W'$, first note that W, ∂W is an integral homology manifold pair (because K, ∂K is). So the constructions of [12] may be applied to equip the cellular chain complex for W, ∂W--denoted $\mathcal{C}$, $\mathcal{C}_\partial \equiv \bigcup_{i=1}^{q} \mathcal{C}_{\partial_i}$--with an algebraic Poincare duality structure. Now complete surgery on the chain complex $\mathcal{C}_{\partial_1}$: note that $\dim(\partial W) = 4\ell - 1$, so computations in [15] show there is no surgery obstruction to completing surgery. Let $(D,\partial_+D,\partial_-D)$ be the surgery cobordism from $\partial_-D = C_{\partial_1}$ to ∂_+D which is an integral homology sphere. Set $W' \equiv C \cup (\underbrace{D \cup D \cup D \cup \ldots \cup D}_{q\text{-fold}})$, where the j-th copy of D is glued to C along the composite $\partial_-D = C_{\partial_1} \to C_{\partial_j}$. Here $C_{\partial_1} \to C_{\partial_j}$ comes from the identifications $\partial_1 W \to \partial_2 W \to \ldots \to \partial_q W$.

This completes the proof of 1.7.

Proof of 1.10. This is proved in [8].

Construction of $\sigma^i(\varphi,R)$.

Recall that $\varphi : Z_p \times R \to R$ is a PL action which has K for fixed point set. We will assume that 1.1 is still in effect.

In constructing $\sigma^i(\varphi,R)$ we follow the same program as for $\gamma^i(R)$. The first step is to construct homomorphisms

1.11. $b_i : \Omega_{4i}(R/\dot{R}) \to W(QZ_p)) \otimes_Z Z_{(2)}$

$b_{i,q} : \Omega_{4i}(R/\dot{R},Z|_q) \to W(Q(Z_p)) \otimes_Z Z_q$

which commute with the change of coefficients $Z \to Z_q$, $Z_q \to Z_{q'}$, (q,q' = powers of 2), and which satisfy $b_i(y \cdot x) = \text{index}(y) \cdot b_i(x)$ or $b_{i,q}(y \cdot x) = \text{index}(y) \cdot b_{i,q}(x)$ for $x \in \Omega_{4i}(R/\dot{R})$ or $x \in \Omega_{4i}(R/\dot{R},Z|_q)$ and $y \in \Omega_*(\text{pt.})$. Then proceeding as in §7 of [13] we can associate to 1.11 a characteristic class $\sigma^i(\varphi,R) \in H^{4i}(R/\dot{R},W(Q(Z_p)) \otimes_Z Z_{(2)})$.

Here is how the $b_i(x)$ is constructed. Represent $x \in \Omega_{4i}(R/\dot{R})$ by $g : N \to R/\dot{R}$ which does not intersect ∂K and is in transverse position to all the simplices of K. By Theorem 1.7 in [10], the equivariant regular neighborhood $\varphi : Z_p \times R \to R$ for K pulls back along $g : g^{-1}(K) \to K$ to a regular neighborhood R' for $g^{-1}(K)$ in N which is equipped with a pulled back action $\varphi' : Z_p \times R' \to R'$ having $g^{-1}(K)$ for fixed point set. Consider the problem of finding an oriented PL cobordism W, W_∂ from $\overline{N - R'}$, R' to X, ∂X where $\dot{R}'$ is the boundary of

R', W_∂ is the product cobordism $\dot{R}' \times [0,1]$, and $\varphi' : Z_p \times (R' \times 1) \to R' \times 1$ extends to a free PL action $\Phi : Z_p \times X \to X$. The obstruction to doing this lies in the relative homotopy group $\pi_{4i+k}(MSPL(k), MSPL(k) \wedge K(Z_p,1))$, where $\{MSPL(k)\}$ is the Thom spectrum for oriented PL bordism theory, and $K(Z_p,1)$ is the Eilenberg-McLane space with $\pi_1 \cong Z_p$. Since this group has order equal to a power of p, say p^s, it follows (after some additional geometric arguments left to the reader) that after replacing the original $x \in \Omega_{4i}(R/\dot{R})$ by $x' = p^s \cdot x$, and letting $g : N \to R/\dot{R}$ represent x' instead of x, that there does exist such a cobordism W, W_∂. Let $\Phi' : Z_p \times N' \to N'$ be the union of $\Phi : Z_p \times X \to X$ and $\varphi' : Z_p \times R' \times 1 \to R' \times 1$ along $(\varphi' : Z_p \times \dot{R}' \times 1 \to \dot{R}' \times 1) = (\Phi : Z_p \times \partial X \to \partial X)$. Now $b_i(x')$ is defined to be the element in $W(Q(Z_p))$ represented by the equivariant intersection form

$$H_{2i}(N',Q) \otimes H_{2i}(N',Q) \to Q(Z_p)\ .$$

Set $b_i(x) \equiv 1/p^s \cdot b_i(x')$.

If $x \in \Omega_{4i}(R/\dot{R}, Z|_q)$ then $b_{i,q}(x) \in W(Q(Z_p)) \otimes_Z Z_q$ is defined similarly to b_i in the preceding paragraph. The only major difference is that the action $\Phi : Z_p \times N' \to N'$ is not now defined on a PL manifold but rather N' is gotten from an oriented PL manifold W with q boundary components $\partial_1 W, \partial_2 W, \ldots, \partial_q W$ by identifying the boundary components by orientation preserving PL homeomorphism $\partial_1 W \to \partial_2 W \to \ldots \to \partial_q W$. And $\Phi' : Z_p \times N' \to N'$ is obtained from a PL action $\zeta : Z_p \times W \to W$, which leaves the $\partial_i W$ invariant and commutes with the identifications $\partial_i W \to \partial_j W$. An element $a_{x'} \in W(Q(Z_p))$ is determined from the equivariant intersection pairing

$$\lambda : H_{2\ell}(W,Q) \times H_{2\ell}(W,Q) \to Q(Z_p)$$

as follows: Define $A \equiv \{y \in H_{2\ell}(W,Q) \mid \lambda(y,y') = 0\ \forall y' \in H_{2\ell}(W,Q)\}$. Then

$$\lambda : \frac{H_{2\ell}(W,Q)}{A} \times \frac{H_{2\ell}(W,Q)}{A} \to Q(Z_p)$$

is non-singular, so it represents an element $a_{x'} \in W(Q(Z_p))$. Arguing as in §1 of [13], we see that, modulo elements of $W(Q(Z_p))$ divisible by q, $a_{x'}$ depends only on the $Z|_q$ cobordism class of $g : N \to R/\dot{R}$. [Note: what precisely is needed here to extend the arguments of §1 in [13] to the present case is an extension of Theorem 1.2 in [13]--called Novikov's additivity theorem--to cover the cases when (in the notation of [13]) M_1^n, M_2^n are oriented PL manifolds equipped with PL actions $\varphi_i \cdot Z \times M_i^n \to M_i^n$ and $I(M_i^n)$ is the element of $W(Q(Z_p))$ represented by the equivariant intersection pairing

$$\lambda : \frac{H_{n/2}(M_j^n,Q)}{A_i} \times \frac{H_{n/2}(M_i^n,Q)}{A_i} \to Q(Z_p)$$

where $A_i \equiv \{x \in H_{n/2}(M_i^n,Q) \mid \lambda(x,y) = 0 \; \forall y \in H_{n/2}(M_i^n,Q)\}$.] Now defining $b_{i,q}(x') \equiv$ image of $a_{x'}$ in $W(Q(Z_p)) \otimes_Z Z_q$, and $b_{i,q}(x) \equiv 1/p^s \cdot b_{i,q}(x')$, well defines $b_{i,q}$. This completes the construction of the $\sigma^i(\varphi,R)$.

Section 2. In this section a relationship between the two characteristic classes $\sum_i \gamma^i$ and $\sum_i \sigma^i$ is described and proven. The proof is just an easy application of the PL equivariant index theorem of [9]. So we begin by reviewing this index theorem.

W, ∂W denotes an orient PL manifold pair of dimension 4ℓ; $\partial W = \bigcup_{i=1}^q \partial_i W$ where the $\partial_i W$ are disjoint components of ∂W identified by orientation preserving PL homeomorphisms $\partial_1 W \to \partial_2 W \to \partial_3 W \to \dots \to \partial_q W$. $\varphi : Z_p \times W \to W$ is a PL action mapping each $\partial_i W$ to itself and commuting with the identifications $\partial_1 W \to \partial_2 W \to \dots \to \partial_q W$. X is the fixed point set of φ, $\{X_j\}$ the connected components of X, $\partial_i X_j \equiv X_j \cap \partial_i W$, $\partial X_j = \bigcup_i \partial_i X_j$. Note that each X_j, ∂X_j is a Z_p-homology manifold pair, so it is also a Q-homology manifold pair. Write $\dim(X_j) = k_j$: We assume in all that follows that $H_{k_j}(X_j/\partial X_j, Z) \cong Z$, i.e., that X_j, ∂X_j is Q-orientable.

The Invariants $i(W)$, $i_\eta(W) \in Z$.

2.0. A hermitian form over $Q(Z_p)$ consists of an f.g. $Q(Z_p)$-module V, and a bilinear map $\lambda : V \times V \to Q(Z_p)$ satisfying:

(i) $\lambda(a \cdot x, \beta \cdot y) = a \cdot \bar{\beta}\lambda(x,y)$, $\forall a,\beta \in Q(Z_p)$ and $x, y \in V$, where if t is the generator for Z_p and $\beta = \sum_{i=0}^{p-1} a_i t^i$, then $\bar{\beta} \equiv a_0 + \sum_{i=1}^{p-1} a_{p-i} t^i$.

(ii) $\det(\hat{\lambda}) \neq 0$ where $\hat{\lambda} : V \times V \to Q$ is defined by $\hat{\lambda}(x,y) = a_0 \Longleftrightarrow \lambda(x,y) = a_0 + \sum_{i=1}^{p-1} a_i t^i$.

Now given (λ, V), define $i(\lambda,V)$, $i_\eta(\lambda,V)$ as follows. Set $V_\eta \equiv \{x \in V \mid \eta \cdot x = 0\}$ where $\eta \equiv \sum_{i=0}^{p-1} t^i$ is the "norm" element of $Q(Z_p)$. $\hat{\lambda}_\eta : V_\eta \times V_\eta \to Q$ is just the restriction to $V_\eta \times V_\eta$ of $\hat{\lambda}$. Note that $\det(\hat{\lambda}_\eta) \neq 0$. Set $i_\eta(\lambda,V) \equiv \mathrm{index}(\hat{\lambda}_\eta)$, and $i(\lambda,V) \equiv \mathrm{index}(\hat{\lambda})$. Now to define $i(W)$, $i_\eta(W)$, let $\lambda : H_{2\ell}(W,Q) \times H_{2\ell}(W,Q) \to Q(Z_p)$ denote the intersection form for W, and $A \equiv \{x \in H_{2\ell}(W,Q) \mid \lambda(x,y) = 0 \ \forall y \in H_{2\ell}(W,Q)\}$. Then

$$\lambda : \frac{H_{2\ell}(W,Q)}{A} \times \frac{H_{2\ell}(W,Q)}{A} \to Q(Z_p)$$

is a hermitian form. Set $i(W) \equiv i(\lambda,V)$, $i_\eta(W) \equiv i_\eta(\lambda,V)$ where $V = \dfrac{H_{2\ell}(W,Q)}{A}$.

<u>The Invariant</u> $i_p(X) \in W(Z_p)$.

$W(Z_p)$ is the Witt group of non-singular symmetric forms over the field Z_p. For any odd prime p $W(Z_p)$ is isomorphic to one of Z_4, $Z_2 \oplus Z_2$.

Choose arbitrarily an integral orientation class $[X_j]$ for X_j. If dimension$(X_j) \neq 0 \bmod 4$, set $i_p(X_j) = 0$. If dimension$(X_j) = 0 \bmod 4$, write $\dim(X_j) = 4\ell_j$, and define $a_j \in W(Z_p)$ to be the element represented by

$$\lambda_j : \frac{H_{2\ell_j}(X_j,Z_p)}{A_j} \times \frac{H_{2\ell_j}(X_j,Z_p)}{A_j} \to Z_p,$$

which is determined by the orientation $[X_j]$. Here $A_j \equiv \{x \in H_{2\ell_j}(X_j,Z_p) \mid \lambda_j(x,y) = 0 \ \forall y \in H_{2\ell_j}(X_j,Z_p)\}$. Set $i_p(X_j) \equiv h_j(a_j)$, and define $i_p(X) \equiv \sum_j i_p(X_j)$.

It still remains to define the homomorphism $h_j : W(Z_p) \to W(Z_p)$. To do this, choose a "slice" of the action $\varphi : Z_p \times W \to W$ near the fixed point component X_j. This will be a free PL action $\varphi_j : Z_p \times S^{4(\ell-\ell_j)-1} \to S^{4(\ell-\ell_j)-1}$ obtained by first constructing an equivariant cell subdivision of $\varphi : Z_p \times W \to W$ "dual" to an equivariant triangulation of $\varphi : Z_p \times W \to W$, and then letting φ_j be the restriction of φ to the boundary of a cell $D^{4(\ell-\ell_j)}$ which is dual to a $4\ell_j$-dimensional simplex of X_j. Note that the orbit space $L_j \equiv \frac{S^{4(\ell-\ell_j)-1}}{Z_p}$ is a homotopy lens space. $[X_j]$ and $[W]$ determine an orientation $[L_j]$ for L_j by requiring that locally the formula $[W] = [X_j] \times [\hat{L}_j]$ must hold $[L_j]$ determines the linking pairing

$$\text{link} : H_{2(\ell-\ell_j)-1}(L_j, Z) \times H_{2(\ell-\ell_j)-1}(L_j, Z) \to Q/Z .$$

Set h_j = identity if for some $x \neq 0$, $\text{link}(x,x) = a/p$ where the integer a is a square mod p. Otherwise, define h_j to send a symmetric matrix $[a_{ij}]$ to $[b \cdot a_{ij}]$, where b is a non-square in Z_p.

<u>Theorem 2.1</u> (PL <u>equivariant index theorem</u>). <u>The invariants</u> $i_p(X)$, $i(W)$, $i_\eta(W)$ <u>are related as in the following tables</u>. $d(q,8)$ <u>denotes the largest common factor of</u> q <u>and</u> 8.

TABLE (a)

If $p = 2r + 1$ with r = odd :

$i_p(X)$	$i_\eta(W)$ mod $d(q,8)$
$[1]$	$-(p-1) \cdot i(W) + 2r$
$[1] \oplus [1]$	$-(p-1) \cdot i(W) + 4$
$[-1]$	$-(p-1) \cdot i(W) - 2r$
$\begin{bmatrix} 0 & 1 \\ 1 & 0 \end{bmatrix}$	$-(p-1) i(W)$

TABLE (b)

If $p = 4r + 1$ with $r = $ odd :

$i_p(X)$	$i_\eta(W) \bmod d(q,8)$
[1]	$-(p-1)\cdot i(W) + 4$
[2]	$-(p-1)i(W)$
$[1] \oplus [2]$	$-(p-1)i(W) + 4$
$\begin{bmatrix} 0 & 1 \\ 1 & 0 \end{bmatrix}$	$-(p-1)\cdot i(W)$

TABLE (c)

If $p = 8r + 1$.

$i_p(X)$	$i_\eta(W) \bmod d(q,8)$
[1]	0
[ℓ]	4
$[1] \oplus [\ell]$	4
$\begin{bmatrix} 0 & 1 \\ 1 & 0 \end{bmatrix}$	0

Remark. The ℓ listed in Table (c) can be any non-square element of Z_p . The symmetric matrices listed under $i_p(X)$ in each of the Tables (a), (b), (c) exhaust all the elements of $W(Z_p)$ for each of the three types of p .

Proof of Theorem 2.1. The reader is referred to [9].

Now we shall discuss the relationship between the characteristic classes $\sum_i \sigma^i$, $\sum_i \gamma^i$.

$\varphi : Z_p \times R \to R$, $K \subset R$, $\gamma^i(R) \in H^{4i}(R/\dot{R}, W(Q) \otimes_Z Z_{(2)})$ and $\sigma^i(\varphi,R) \in H^{4i}(R/\dot{R}, W(Q(Z_p)) \otimes_Z Z_{(2)})$ are all as in Section 1 above.

2.2. Construction of $\bar{\sigma}^i(\varphi,R) \in H^{4i}(R/\dot{R},Z_8)$. Define a homomorphism $g : W(Q(Z_p)) \to Z_8$ as follows. Represent $x \in W(Q(Z_p))$ by a hermitian form $\lambda : V \times V \to Q(Z_p)$. Set $g(x) \equiv i_\eta(\hat{\lambda}) + (p-1)i(\hat{\lambda}) \bmod 8$ (see beginning of §2 for i , i_η) . Let $G_i : H^{4i}(R/\dot{R},W(Q(Z_p)) \otimes_Z Z_{(2)}) \to H^{4i}(R/\dot{R},Z_8)$ be the map induced by the coefficient map $g \otimes 1_{Z_{(2)}}$. Define $\bar{\sigma}^i(\varphi,R) \equiv G_i(\sigma^i(\varphi,R))$.

2.3. Construction of $\bar{\gamma}^i(R) \in H^{4i}(R/\dot{R},Z_8)$. $[K]$, $[R]$ are the given integral homology orientation classes for K, R. Let L denote the homotopy lens space associated to a "slice" of the action $\varphi : Z_p \times R \to R$ near K (see the construction of $i_p(X)$ above). And let $[L]$ be the integral orientation class for L consistent with $[K]$ and $[R]$. $[L]$ determines a linking pairing

$$\text{link} : H_{2(r-k)-1}(L,Z) \times H_{2(r-k)-1}(L,Z) \to Q/Z ,$$

where $\dim(K) = 4k$, $\dim(R) = 4r$. Define an invariant $O(\varphi) \in Z_2$ to be 1 if $\exists x \in H_{2(r-k)}(L,Z)$ with $\text{link}(x,x) = a/p$ where the integer a is a non-square mod p, and otherwise set $O(\varphi)$ equal to zero.

Now note that since K is the fixed set for $\varphi : Z_p \times R \to R$, K, ∂K must be a $Z_{(p)}$-homology manifold pair. So each $\gamma^i(R)$ pulls back along the inclusion of coefficients $W(Z_{(p)}) \otimes_Z Z_{(2)} \subset W(Q) \otimes_Z Z_{(2)}$ to a class $\gamma_1^i(R) \in H^{4i}(R/\dot{R},W(Z_{(p)}) \otimes_Z Z_{(2)})$.

Define $f_1 : W(Z_{(p)}) \to W(Z_p)$ to be reduction mod p.

Define $f_2 : W(Z_p) \to Z_8$ as follows.

(a) If $p = 2r + 1$, r = odd; then $W(Z_p) \cong Z_4$ and the matrix $[1]$ is a generator. Set $f_2(n \cdot [1]) = n \cdot 2r \bmod 8$ if $O(\varphi) = 0$. And set $f_2(n \cdot [1]) = -n \cdot 2r \bmod 8$ if $O(\varphi) = 1$.

(b) If $p = 4r + 1$, r = odd; then $W(Z_p) \cong Z_2 \oplus Z_2$ with generators $[1], [2]$ respectively. Set $f_2(n[1] + m[2]) \equiv n \cdot 4 \bmod 8$ if $O(\varphi) = 0$. And set $f_2(n[1] + m[2]) \equiv m \cdot 4 \bmod$ if $O(\varphi) = 1$.

(c) If $p = 8q + 1$; then $W(Z_p) \cong Z_2 \oplus Z_2$ with generators $[1]$, $[\ell]$, where ℓ is not a square mod p. Set $f_2(n \cdot [1] + m[\ell]) \equiv m \cdot 4 \bmod 8$ if $O(\varphi) = 0$. And set $f_2(n \cdot [1] + m[\ell]) = n \cdot 4 \bmod 8$ if $O(\varphi) = 1$.

Finally, set

$$F_i : H^{4i}(R/\dot{R},W(Z_{(p)}) \otimes_Z Z_{(2)}) \to H^{4i}(R/\dot{R},Z_8)$$

be the map induced by the coefficient map $(f_2 \cdot f_1) \otimes 1_{Z_2} : W(Z_{(p)}) \otimes_Z Z_{(2)} \to Z_8$. Define $\overline{\gamma}^i(R) \equiv F_i(\gamma_1^i(R))$.

<u>Theorem 2.4</u>. $\sum_i \overline{\gamma}^i(R) = \sum_i \overline{\sigma}^i(\varphi,R)$.

<u>Remark</u>. 2.4 is the characteristic class version of the PL equivariant index Theorem 2.1.

<u>Proof of Theorem 2.4</u>. <u>Step 1</u>. First note that the constructions of the type in §7 of [13] associate to the homomorphisms $\{f_2 \cdot f_1 \cdot a_i, (f_2 \cdot f_1 \otimes 1_{Z_q}) \cdot a_{i,q}\}$ and $\{(g \times 1_{Z_{(2)}}) \cdot b_i, g \otimes 1_{Z_q} \cdot b_{i,q}\}$ precisely the classes $\overline{\gamma}^i(R)$ and $\overline{\sigma}^i(R)$. Here g is from 2.2; f_1, f_2 are from 2.3; $\{a_i, a_{i,q}\}$ are from 1.2, 1.3; $\{b_i, b_{iq}\}$ are from 1.11. Although in 1.2, 1.3 the ranges of a_i, $a_{i,q}$ are $W(Q)$ and $W(Q) \otimes Z_q$, because K, ∂K is a $Z_{(p)}$-homology manifold pair, the images of a_i, $a_{i,q}$ actually lie in the subgroups $W(Z_{(p)})$ and $W(Z_{(p)}) \otimes Z_q$. So f_1 and $f_1 \otimes 1_{Z_q}$ can be composed with the a_i, $a_{i,q}$.

<u>Step 2</u>. Next note that the $\{f_2 \cdot f_1 \otimes 1_{Z_{(2)}} \cdot a_i, f_2 \cdot f_1 \otimes 1_{Z_q} \cdot a_{i,q}\}$ are equal to the $\{g \otimes 1_{Z_{(2)}} \cdot b_i, g \otimes 1_{Z_q} \cdot b_{i,q}\}$. This is an immediate consequence of Theorem 2.1 and the construction of these maps. Because the constructions of the type in §7 of [13] associate to identical sets of maps, identical cohomology classes, we get by Step 1 that $\overline{\gamma}^i(R) = \overline{\sigma}^i(R) \forall i$.

This completes the proof of 2.4.

<u>Section 3</u>. In this section Proposition 0.1 of the introduction is proved.

<u>Construction of</u> $\sum_i h_i^p(K) \in \sum_i H_{4i-1}((K,\partial K),Z)$.

Denote dimension$(K) = 4k$. Choose $s \gg k$, embed K, ∂K in $(B^{4s}, \partial B^{4s})$ so that K intersects ∂B^{4s} transversely in ∂K. R, R_∂ denotes a PL regular neighborhood for K, ∂K in B^{4s}, ∂B^{4s}. Here B^{4s} is the 4s-dimensional ball. Under the composition of the Poincaré duality map $H^{4i}(R/\dot{R}, Z_8) \cong H_{4(s-i)}(R/R_\partial, Z_8)$ with the homotopy retraction map $H_{4(s-i)}(R/R_\partial, Z_8) \cong H_{4(s-i)}((K,\partial K, Z_8)$ the class

$\bar{\gamma}^i(R)$ of 2.3 above goes to a homology class $\bar{\gamma}_{s-i}(R) \in H_{4(s-i)}((K,\partial K),Z_8)$. Define $h^p_{s-i}(K)$ to be the image of $\bar{\gamma}_{s-i}(R)$ under the Bockstein map

$$H_{4(s-i)}((K,\partial K),Z_8) \xrightarrow{\beta_{4(s-i)}} H_{4(s-i)-1}((K,\partial K),Z).$$

It is easily checked that the characteristic class $\sum_i h^p_i(K) \in \sum_i H_{4i-1}((K,\partial K),Z)$ depends only on the oriented PL homeomorphism type of K, ∂K.

We will need the following lemma to prove Proposition 0.1.

Lemma 3.1. Given any $x \in H^{4i}(R/\dot{R},Z_8)$ let $x' \in H_{4(s-i)}((K,\partial K),Z_8)$ be its image under the composition of the Poincare duality map $H^{4i}(R/\dot{R},Z_8) \to H_{4(s-i)}(R/R_\partial,Z_8)$ with the homotopy retraction map $H_{4(s-i)}(R/R_\partial,Z_8) \to H_{4(s-i)}((K,\partial K),Z_8)$. Then $\beta_{4(s-i)}(x') = 0$ if and only if the cap product $x \cap : H_{4i}(R/\dot{R}) \to Z_8$ vanishes on the torsion subgroup of $H_{4i}(R/\dot{R})$.

Proof of 3.1. Let $x'' \in H_{4(s-i)}(R/R_\partial,Z_8)$ be the image of x under Poincaré duality. By using the geometric form of Poincaré duality (intersection and linking pairing) one sees that $\beta_{4(s-i)}(x'') = 0 \Longleftrightarrow x \cap : H_{4i}(R/\dot{R}) \to Z_8$ vanishes on the torsion subgroup. But $\beta_{4(s-i)}(x'') = 0 \Longleftrightarrow \beta_{4(s-i)}(x') = 0$.

This completes the proof of 3.1.

Proof of Proposition 0.1.

Part (a). By 3.1 it suffices to show that the cap product $(2 \cdot \bar{\gamma}^{4j}(R)) \cap :$ $H_{4j}(R/\dot{R}) \to Z_8$ vanishes on the torsion subgroup $\forall j$. First, note that by construction the cap product map $\bar{\gamma}^{4j}(R) \cap$ tensored with $1_{Z_{(2)}}$ equals the restriction of $(f_2 \cdot f_1 \cdot a_j) \otimes 1_{Z_{(2)}} : \Omega_{4j}(R/\dot{R}) \otimes Z_{(2)} \to Z_8$ to $H_{4j}(R/\dot{R},\Omega_0) \otimes Z_{(2)}$ under the equivalence of homology functors

$$\Omega_j(\ ) \otimes Z_{(2)} \cong \sum_j H_{j-k}(\ ,\Omega_k) \otimes Z_{(2)}.$$

For details see 2.3, 1.2 above and §7 of [13]. So it suffices to show that

$2\cdot(f_2\cdot f_1\cdot a_j)$ vanishes on the torsion subgroup of $\Omega_{4j}(R/\dot{R})\ \forall j$. If $x\in\Omega_{4j}(R/\dot{R})$ is in the torsion subgroup and $h: N\to R/\dot{R}$ is a map from a differentiable manifold representing x, which is in transverse position to all simplices of K, then $h^{-1}(K)$ is an oriented $Z_{(p)}$-homology manifold with <u>vanishing index</u> (because x is torsion). Hence the mod p reduction of the $Z_{(p)}$-homology middle dimension intersection form for $h^{-1}(K)$ has even rank and therefore has order at most 2 in $W(Z_p)$. So $2\cdot(f_2\cdot f_1\cdot a_j)=0$.

<u>Part</u> (b). Choose an equivariant PL regular neighborhood $\varphi: Z_p\times R', R'_\partial\to R'$, R'_∂ for K, ∂K in $\varphi: Z_p\times M$, $\partial M\to M$, ∂M. First consider the case $R'=R$, where R is the regular neighborhood for K, ∂K in B^{4s}, ∂B^{4s}.

Let $b'_j:\Omega_{4j}(R/\dot{R})\to Z$ be the composite of $b_j:\Omega_{4j}(R/\dot{R})\ \Omega\ W(Q(Z_p))\otimes_Z Z_{(2)}$ in 1.11 with $(i+(p-1)i_\eta)\otimes 1_{Z_2}: W(Q(Z_p))\otimes_Z Z_{(2)}\to Z_{(2)}$ in 2.2. b'_j vanishes on the torsion subgroup because image (b'_j) is torsion free. So $(g\otimes 1_{Z_{(2)}})\cdot b_j$ vanishes on the torsion subgroup, where g comes from 2.2. The cap product of $\sigma^j(\varphi,R)\cap: H_{4j}(R/\dot{R})\to Z_8$ tensored with $1_{Z_{(2)}}$ is equal to the restriction of

$$(g\otimes 1_{Z_{(2)}})\cdot(b_j\otimes 1_{Z_{(2)}}):\Omega_{4j}(R/\dot{R})\otimes Z_{(2)}\to Z_8$$

to the subgroup $H_{4j}(R/\dot{R},\Omega_0)\otimes Z_{(2)}$ under the equivalence $\Omega_{4j}(R/\dot{R})\otimes Z_{(2)}\cong \sum_k H_{4j-k}(R/\dot{R},\Omega_k)\otimes Z_{(2)}$. So $\bar{\sigma}^j(\varphi,R)\cap$ vanishes on the torsion subgroup of $H_{4j}(R/\dot{R})$. But $\bar{\sigma}^j(\varphi,R)=\bar{\gamma}^j(R)$ by 2.4. So by 3.1 $h^p_{4(s-j)}=0\ \forall j$.

This completes the proof of Part (b) if $R=R'$. In general note that a_j, b_j extend to mappings on PL bordism groups

$$(a_j,b_j):\Omega^{PL}_{4j}(R/\dot{R})\to (W(Q),W(Q(Z_p))\otimes_Z Z_{(2)})$$

by using the same geometric definitions as in §1. There are commutative diagrams

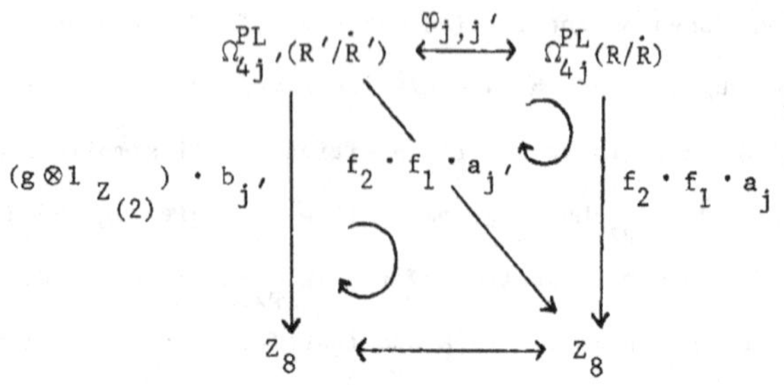

Here $4(j-j') = \dim(R)-\dim(R')$, $\varphi_{j,j'}$ is the Thom isomorphism, f_1 , f_2 are as in 2.3, g is as in 2.2, $a_{j'}$, a_j come from 1.2 and $b_j{}'$ from 1.11. The diagram to the left is commutative by 2.4. The diagram to the right is commutative because $\varphi_{j,j'}$ preserves transversal intersections. In the preceding paragraph, we showed that $(g \otimes 1_{Z_{(2)}}) \cdot b_{j'}$ vanished on the torsion subgroup. Hence, by commutativity of the above diagrams $f_2 \cdot f_1 \cdot a_j$ vanishes on the torsion subgroup. But then, arguing as in the preceding paragraph, the cap product $\overline{\gamma}^j(R) \cap : H_{4j}(R/\dot{R}) \to Z_8$ vanishes on the torsion subgroup. So by 3.1, $h^P_{4(s-j)} = 0 \ \forall j$.

Part (c). First note that for $f_2 \cdot f_1$ as in 2.3, and a_j , $a_{j,T}$ as in 1.2, 1.5, we have $f_2 \cdot f_1 \cdot a_{j,T} = f_2 \cdot f_1 \cdot a_j$ holds on the torsion subgroup. Because K , ∂K is an integral homology manifold pair $a_{j,T} = 0$ (see 1.4, 1.5, 1.7). So $f_2 \cdot f_1 \cdot a_j = 0$ on the torsion subgroup. Since the cap product $\overline{\gamma}_j(R) \cap$: $H_{4j}(R/\dot{R}) \to Z_8$ tensored with $1_{Z_{(2)}}$ is equal to the restriction of

$$(f_2 \cdot f_1 \cdot a_j) \otimes 1_{Z_{(2)}} : \Omega_{4j}(R/\dot{R}) \otimes Z_{(2)} \to Z_8$$

to the subgroup $H_{4j}(R/\dot{R},\Omega_0) \otimes Z_{(2)}$ under the equivalence $\Omega_{4j}(R/\dot{R}) \otimes Z_{(2)} = \sum_k H_{4j-k}(R/\dot{R},\Omega_k) \otimes Z_{(2)}$, we get that $\overline{\gamma}_j(R) \cap$ vanishes on the torsion subgroup of $H_{4j}(R/\dot{R})$. So $h^P_{4(s-j)} = 0 \ \forall j$ by 3.1.

Part (d). This follows immediately from 1.10.

Part (e). $[m_{ij}]$ denotes a symmetric matrix over the integers satisfying: $m_{ii}/2$ = integer $\forall i$; $\det([m_{ij}])$ is a unit mod p ; index $([m_{ij}]) = 0$; the mod p reduction of $[m_{ij}]$ is not zero in $W(Z_p)$. For example, if $p = 2r+1$, r = odd , then $\begin{bmatrix} 4 & 0 \\ 0 & -4 \end{bmatrix} = [m_{ij}]$ will do. Or if $p = 4r+1$, r = odd , then $\begin{bmatrix} 2 & 0 \\ 0 & -4 \end{bmatrix} = [m_{ij}]$ will work.

Using "plumbing constructions" extend D^8 to the framed manifold W^8 satisfying:

i) W^8 is 3-connected;

ii) the 4-dimensional intersection pairing on W^8 realizes $[m_{ij}]$. Set $P_1 \equiv W^8 \cup \text{cone}(\partial W^8)$. ∂W^8 is a $Z_{(p)}$-homology sphere, so P_1 is a $Z_{(p)}$-homology manifold. In $P_1 \times S^\ell$ $(\ell \geq 6)$ one can find four disjoint copies of P_1; let X denote their connected sum. The intersection form for X, $[m_{ij}] \oplus [m_{ij}] \oplus [m_{ij}] \oplus [m_{ij}]$, is congruent over $Z_{(p)}$ to a sum of hyperbolic planes because $\text{index}([m_{ij}]) = 0$ and $W_T(Z_{(p)})$ has exponent equal to 4. Thus, if V is a maximal subspace of $H_4(X, Z_{(p)})$ which is perpendicular to itself, then $2 \dim_{Z_{(p)}}(V) = \dim_{Z_{(p)}}(H_4(X,Q))$. Let $(N,\bar{X})$ denote the cobordism of $(P_1 \times S^\ell, X)$ obtained by performing framed surgeries, along framed spheres in the homogeneous part of X, which represent a $Z_{(p)}$ basis for V. A calculation shows $\partial_+\bar{X}$ is a $Z_{(p)}$-homology 8-sphere. Next perform framed surgeries in the homogeneous part of $\partial_+N - \partial_+\bar{X}$ to kill both $H_4(\partial_+N, Z_{(p)})$, $H_5(\partial_+N, Z_{(p)})$ (note: $\ell \geq 6 \Longrightarrow \text{codim}_N(\bar{X}) > 5$). This can be accomplished without adding any further classes to $H_6(\partial_+N, Z_{(p)})$. Now a calculation shows $\bar{H}_*(\partial_+N, Z_{(p)})$ is generated by ∂_+N, S^ℓ, $\partial_+\bar{X}$, the latter two of these being dually paired. Finally, noting that $D^\ell \times \bar{X}$ is a regular neighborhood for $\bar{X}$ in N, extend N to $\bar{N} = N \cup (D^\ell \times \text{cone}(\partial_+\bar{X}))$. Set K equal the complement in $\partial_+\bar{N}$ of an open $(\ell+8)$-ball. Then K is a $Z_{(p)}$-homology manifold; K is a $Z_{(p)}$-homology disc; P_1 is contained in K with $P_1 \times D^\ell$ for regular neighborhood. P_1 represents the transversal intersection with K of a torsion bordism element $h : N \to R/\dot{R}$, where R, R_∂ is a regular neighborhood for a PL embedding of K, ∂K in a ball pair B^{4s}, ∂B^{4s} with $s \gg \dim(K)$. Because the middle dimensional intersection form for $h^{-1}(K)$ is therefore represented by $[m_{ij}]$, which is non-zero in

$W(Z_p)$, it follows that not all of the maps $f_2 \cdot f_1 \cdot a_j : \Omega_{4j}(R/\dot{R}) \to Z_8$ of 1.2, 2.3 can vanish on the torsion subgroup. So not all of the restrictions of $(f_2 \cdot f_1 \cdot a_j) \otimes 1_{Z_{(2)}}$, to $H_{4j}(R/\dot{R},\Omega_0) \otimes Z_{(2)}$ can vanish on the torsion subgroups. These restrictions are equal to the cap product $\bar{\gamma}^j(R) \cap : H_{4j}(R/\dot{R}) \to Z_8$ tensored with 1_{Z_2}, so by 3.1 not all of the $h^p_{4(s-j)}(K)$ can vanish.

This completes the proof of 0.1.

[1] Alexander, J.; Hammerick, G.; Vick, J., Bilinear forms and cyclic group actions, Bull. Amer. Math. Soc. 80 (1974), 730-734.

[2] Alexander, J.; Hammerick, G., Periodic maps on Poincaré spaces, preprint.

[3] Conner, P.E.; Raymond, F., A quadratic form on the quotient of a periodic map, Semi-group Forum, Springer-Verlag 7, 310-333.

[4] Husemoller, D.; Milnor, J., Symmetric Bilinear Forms, Springer-Verlag, 1973.

[5] Jones, L., Combinatorial symmetries of the m-disc, preprint, Berkeley, 1972.

[6] ______, Notices of American Math. Soc. 19 (1972), A-545.

[7] ______, Combinatorial symmetries of the m-disc, Bull. Amer. Math. Soc. 79 (1973), 167-169.

[8] ______, Topological invariance of certain combinatorial characteristic classes, Bull. Amer. Math. Soc. 79 (1973), 981-983.

[9] ______, A relation between indexes, submitted to Compositio Mathematica (1976).

[10] ______, Geometric constructions for $Z_{(n)}$-homology manifolds, to appear in Proceedings of London Math. Soc.

[11] Milnor, J., On simply connected 4-manifolds, Symposium Internacional Topologia Algebraica, Mexico, 1958.

[12] Miščenko, A., Homotopy invariants of non-simply connected manifolds III, Mathematics of the U.S.S.R. 5 (1971), 1325-1364.

[13] Morgan, J.; Sullivan, D., The transversality characteristic class and linking cycles in surgery theory, Annals of Math. 99 (1974), 463-544.

[14] Novikov, Pontryagin classes, the fundamental group, some problems in stable algebra, Essays of Topology and Related Topics, Memoires dediés à George de Rham, Springer-Verlag, 1970.

[15] Ranicki, A., The algebraic theory of surgery, preprint.

[16] Smith, P.A., Transformations of finite period, I, II, Annals of Math. 39 (1938), 127-164, and 40 (1939), 690-711.

THE DEGREE OF MULTIPLE-VALUED MAPS OF SPHERES

Charles N. Maxwell
Southern Illinois University

In this article we shall generalize the concept of degree of a self-map on a sphere to the case where the map is multivalued. More precisely, the functions considered are maps from S^q to $(S^q)^{(n)}$, the n-fold symmetric product of the q-sphere, for some $n > 0$, $q > 0$. Such a map will be called a multimap of multiplicity n on S^q. We will be interested in the homotopy class of such maps, and in section 1 an equivalence relation is introduced on the set of multimaps in an appropriate way so that we may say that two multimaps of perhaps different multiplicities are "homotopic" (see symmetric equivalence (1.8)).

To each multimap f on S^q, we associate a rational number $\deg f$. This number is invariant under symmetric equivalence and we may therefore speak also of $\deg[f]$ of the multimap class $[f]$.

Many of the properties of the classical notion of degree of maps of spheres (see [2], p. 62) prevail in this more general setting. Our main results are:

(i) $\operatorname{Deg} f = 1$ if f is the identity map on S^q.

(ii) If $f: S^q \to S^q$ (that is, f maps into the 1-fold symmetric product), then $\deg[f]$ is an integer which agrees with the number $\deg f$ given by the classical notion of degree.

A composition $[f] * [g]$ of two multimap classes and the suspension $\overline{\overline{\Sigma}}[f]$ of a multimap class will be introduced in section 1. The following are valid:

(iii) $\operatorname{Deg}[f] * [g] = \deg[f] \cdot \deg[g]$.

(iv) $\operatorname{Deg} \overline{\overline{\Sigma}}[f] = \deg[f]$.

(v) For any integer $q > 0$ and rational number r, there is a multimap f on S^q with $\deg[f] = r$.

A multimap f is said to have a fixed point if for some x, $f(x)$ has x as a coordinate.

(vi) For a multimap f on the q-sphere, if q is odd and $\deg[f] \neq 1$, then f has a fixed point, and if q is even and $\deg[f] \neq -1$, then f has a fixed point.

A join $g \vee h: X \to Y^{(n+m)}$ of two multimaps $g: X \to Y^{(n)}$ and $h: X \to Y^{(m)}$ will be defined. Included also is the result that

(vii) $\deg(g \vee h) = \dfrac{n}{n+m} \deg g + \dfrac{m}{n+m} \deg h.$

In section 1, some preliminaries and the precise definition of multimap and multimap class are given. In section 2, the simplicial and homological machinery is set up to define degree. In section 3, the seven properties are proven.

§1. Multimaps. For a positive integer n and topological space E, $E^{(n)}$ will denote the n-fold symmetric product of E and is defined

(1.1) $$E^{(n)} = E^n/\Sigma_n$$

where E^n is the n-fold cartesian product of E and Σ_n is the symmetric group of all permutations of the coordinates. The equivalence class of $(x_1,\dots,x_n) \in E^n$ will be denoted by $[x_1,\dots,x_n]$ or by the word $x_1 \cdots x_n$ of commuting letters. Given a map $f: E \to F$ and a positive integer n, there is a map $f^{(n)}: E^{(n)} \to F^{(n)}$ given by

(1.2) $$f^{(n)}[x_1,\dots,x_n] = [f(x_1),\dots,f(x_n)].$$

The n-fold symmetric product $(\ )^{(n)}$ forms a functor on the category TOP of topological spaces and continuous maps. Since $f \sim g$ implies $f^{(n)} \sim g^{(n)}$, this functor induces a functor on the homotopy category TOP_H.

For a space E and positive integers n and m, there is an adjoining map

(1.3) $$\alpha_n^m(E): E^{(n)} \times E^{(m)} \to E^{(n+m)}$$

given by $\alpha_n^m(E)([x_1,\dots,x_n], [y_1,\dots,y_n]) = [x_1,\dots,x_n, y_1,\dots,y_m]$. For all m and n, α_n^m is natural. More generally, we may speak of an adjoining map for the space E

(1.4) $$E^{(n_1)} \times \cdots \times E^{(n_j)} \to E^{(n_1 + \cdots + n_j)}$$

for positive integers $n_1,\ldots,n_j$.

In particular, the adjoining map

$$(E^{(n)})^m = E^{(n)} \times \cdots \times E^{(n)} \overset{\alpha}{\to} E^{(nm)}$$

is equivariant with respect to the permutations of the m coordinates. Hence there is induced a map

(1.5) $$\beta_n^m(E): \quad (E^{(n)})^{(m)} \to E^{(nm)}$$

given by

$$\beta_n^m[[x_1^1,\ldots,x_n^1],\ldots,[x_1^m,\ldots,x_n^m]] = [x_1^1,\ldots,x_n^1,\ldots,x_1^m,\ldots,x_n^m].$$

β_n^m is also natural for all positive integers m and n.

For a space E and for positive integers n and q such that $q = nm$ for some m, the diagonal Δ_m of $E^{(n)}$ in $(E^{(n)})^m$ induces a map

(1.6) $$d_n^q(E): \quad E^{(n)} \to E^{(q)}$$

defined to be the composition

$$E^{(n)} \xrightarrow{\Delta_m} (E^{(n)})^m \xrightarrow{p} (E^{(n)})^{(m)} \xrightarrow{\beta_n^m} E^{(q)}$$

where p is the canonical projection and $m = q/n$. In word notation for an element $x_1 \cdots x_n$ of $E^{(n)}$, $d_n^q(E)x_1 \cdots x_n = x_1^m \cdots x_n^m$. Note that $d_q^s(E) \cdot d_n^q(E) = d_n^s(E)$ whenever n divides q and q divides s.

(1.7) A <u>multimap of multiplicity</u> $n > 0$ from space X to space Y is a map $f: X \to Y^{(n)}$. Multimaps f and g from X to Y of multiplicity n and m, respectively, are <u>symmetrically equivalent</u> if for some common integer multiple q of n and m, $d_n^q(Y)f$ and $d_m^q(Y)g$ are homotopic.

(1.8) LEMMA. <u>Symmetric Equivalence is an equivalence relation on the set of all multimaps from</u> X <u>to</u> Y.

The proof is omitted. Note that for any space X, the identity map 1_X and $d_1^n(X)$ are symmetrically equivalent, since

$$(1.9)\qquad \begin{array}{ccc} X & \xrightarrow{1_X} & X \\ {\scriptstyle d_1^n(X)}\downarrow & & \downarrow{\scriptstyle d_1^n(X)} \\ X^{(n)} & \xrightarrow[d_n^n(X) = 1_{X^{(n)}}]{} & X^{(n)} \end{array} \quad \text{commutes.}$$

Given maps $f\colon X \to Y^{(n)}$ and $g\colon Y \to Z^{(m)}$, define $g * f\colon X \to Z^{(mn)}$ to be the composition

$$(1.10)\qquad X \xrightarrow{f} Y^{(n)} \xrightarrow{g^{(n)}} (Z^{(m)})^{(n)} \xrightarrow{\beta_m^n(Z)} Z^{(mn)}$$

where β_m^n is defined in (1.5).

(1.11) LEMMA. *If* $f\colon X \to Y^{(n)}$, $\bar{f}\colon X \to Y^{(\bar{n})}$, $g\colon Y \to Z^{(m)}$, $\bar{g}\colon Y \to Z^{(\bar{m})}$ *are multimaps with* f *and* g *symmetrically equivalent to* $\bar{f}$ *and* $\bar{g}$ *respectively, then* $g * f$ *is symmetrically equivalent to* $\bar{g} * \bar{f}$.

The proof is routine and will be omitted. A composition of multimap classes $[g]$ and $[f]$ under symmetric equivalence may be defined by $[g] * [f] = [g * f]$.

(1.12) The category whose objects are all topological spaces, morphisms are multimaps, and composition of multimaps given by $*$ in (1.10) will be denoted by M. The quotient category whose morphisms are multimap classes under symmetric equivalence will be denoted by $\bar{M}$.

(1.13) EXAMPLE. Let G be a group and H a subgroup of G whose index in G is n. Let $G/H = \{Y_1, \ldots, Y_n\}$ be the set of left cosets of H indexed arbitrarily. Group multiplication in G induces a left action $G \times (G/H) \to G/H$ given by $g(Y_i) = (gY_i) \in G/H$. Thus multiplication by g permutes G/H.

In addition, suppose that a right action of G on a topological space X is given, $X \times G \to X$; therefore, upon restriction, we have a right action of H on X. There is a map on the orbit spaces

$$X \backslash H \xrightarrow{u} X \backslash G \quad \text{given by} \quad u(xH) = xG.$$

Now define $X \backslash G \xrightarrow{v} (X \backslash H)^{(n)}$ by $v(xG) = [xY_1, \ldots, xY_n]$. Note that if $Y = aH$ for $a \in G$, then $x(aH) = (xa)H \in X \backslash H$. Furthermore, if $xG = \bar{x}G$, then $x = \bar{x}g$ for some $g \in G$, and $[xY_1,\dots,xY_n] = [\bar{x}gY_1,\dots,\bar{x}gY_n] = [\bar{x}Y_1,\dots,\bar{x}Y_n]$; thus v is well defined.

For any $Y_i = a_iH \in G/H$, the map $\varphi_i\colon X \to X\backslash H$ given by $\varphi_i x = xY$ is continuous, since it is multiplication by a_i followed by projection into $X\backslash H$. Hence the map $v\colon X \to (H\backslash H)^n$ with $\bar{v}(x) = (\varphi_1 x,\dots,\varphi_n x)$ is continuous. The continuity of v follows from the commutativity of the following diagram where $\bar{p}$ and p are the projections into the orbit spaces $\operatorname{mod} G$ and Σ_n respectively:

$$\begin{array}{ccc} X & \xrightarrow{\quad \bar{v} \quad} & (X\backslash H)^n \\ {\scriptstyle \bar{p}}\downarrow & & \downarrow{\scriptstyle p} \\ X\backslash G & \xrightarrow{\quad v \quad} & (X\backslash H)^{(n)}. \end{array}$$

Hence, v is a multimap. The composition $[u]*[v] = [1]\colon X\backslash G \to X\backslash G$, since the triangle

$$\begin{array}{ccc} X\backslash G & \xrightarrow{\qquad v \qquad} & (X\backslash H)^{(n)} \\ & \searrow{\scriptstyle d_1^n(X\backslash G)} & \downarrow{\scriptstyle u^{(n)}} \\ & & (X\backslash G)^{(n)} \end{array}$$

commutes, and $d_1^n(X\backslash G)$ is symmetrically equivalent to the identity on $X\backslash G$.

(1.14) EXAMPLE. If $\pi\colon E \to B$ is a finite covering of degree n, with E path connected, let $H = \pi_1(E)$ and $G = \pi_1(B)$. Then H may be identified with a subgroup of G of index n, and the given covering may be identified with the homeomorphic covering $u\colon X\backslash H \to X\backslash G$ (where X is the universal covering space of B and X is acted upon freely on the right by G). Then (1.13) applies to give a map $v\colon B \to E^{(n)}$ so that $[u]*[v] = [1_B]$.

(1.15) EXAMPLE. Specializing further, let $E = B = S^1$ = the complex numbers of norm 1. For an integer a, let $u_a\colon S^1 \to S^1$ be defined by $u_a(Z) = Z^a$. If $a \neq 0$, then u_a is an $|a|$-fold covering map. Let $v_a\colon S^1 \to (S^1)^{(n)}$ be the map obtained in (1.14) with the property that $[u_a]*[v_a] = [1_{S^1}]$. For positive integers a and b, $u_a * v_b$ is the multimap which sends $Z \in S^1$ into the unordered set of $a^{\underline{th}}$ powers of the b $b^{\underline{th}}$ roots of Z. (See (3.6).)

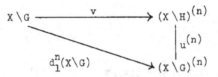

(1.16) Given maps $f\colon X \to Y^{(n)}$ and $g\colon X \to Y^{(m)}$, the <u>join</u> $f \vee G\colon X \to Y^{(n+m)}$ is defined as the composition

$$X \xrightarrow{f \times g} Y^{(n)} \times Y^{(m)} \xrightarrow{\alpha_n^m(Y)} Y^{(n+m)}$$

where α is defined in (1.3).

If f and f' are homotopic maps from X to $Y^{(n)}$, and if g and g' are homotopic maps from X to $Y^{(m)}$, then $f \vee g$ is homotopic to $f' \vee g'$. (The analogous statement for symmetric equivalence is not valid.)

The join is commutative and associative, and by iteration there is defined a join $f_1 \vee \cdots \vee f_k\colon X \to Y^{(n_1 + \cdots + n_k)}$ of maps $f_i\colon X \to Y^{(n_i)}$, $i = 1,\ldots,k$.

(1.17) A suspension functor $\bar{\Sigma}$ may be defined on the categories M and $\bar{M}$ of (1.12). Let ΣX denote the (unreduced) suspension $X \times I/\sim$ of a space X. An element of ΣX will be written $[x,t]$ with $x \in X$, $0 \leq t \leq 1$. There is a natural map

$$\gamma_n(X)\colon \Sigma(X^{(n)}) \to (\Sigma X)^{(n)}$$

given by $\gamma_n(X)[[x_1,\ldots,x_n], t] = [[x_1, t],\ldots,[x_n, t]]$.

Since $q \times 1_I$ is an identification map whenever q is such (see [1], p. 101), one can prove that $\gamma_n(X)$ is well-defined and continuous, by referring to the following diagram where p is the identification map of suspension and q is the identification map of symmetric product.

$$\begin{array}{ccc}
X^n \times I & \xrightarrow{\bar{\gamma}_n(X)} & (X \times I)^n \\
\downarrow{\scriptstyle q \times 1} & & \downarrow{\scriptstyle q} \\
X^{(n)} \times I & & (X \times I)^{(n)} \\
\downarrow{\scriptstyle p} & & \downarrow{\scriptstyle p^{(n)}} \\
\Sigma X^{(n)} & \dashrightarrow{\gamma_n(X)} & (\Sigma X)^{(n)}
\end{array}$$

where $\bar{\gamma}_n(X)(x_1,\ldots,x_n, t) = ((x_1, t),\ldots,(x_n, t))$

Given a map $f\colon X \to Y$, then the following diagram commutes:

(1.18)
$$\begin{array}{ccc} \Sigma(X^{(n)}) & \xrightarrow{\Sigma(f^{(n)})} & \Sigma(Y^{(n)}) \\ \downarrow{\scriptstyle \gamma_n(X)} & & \downarrow{\scriptstyle \gamma_n(Y)} \\ (\Sigma X)^{(n)} & \xrightarrow{(\Sigma f)^{(n)}} & (\Sigma Y)^{(n)} \end{array}$$

Hence $\gamma_n(X)$ is a natural transformation from $\Sigma(-^{(n)})$ to $(\Sigma -)^{(n)}$ on TOP.

(1.19) The suspension $\bar{\Sigma}$ is defined on the category M of (1.12) in this way: For an object X, $\bar{\Sigma} X = \Sigma X$. For a multimap $g: X \to Y^{(n)}$ from X to Y, $\bar{\Sigma} g$ is the composition

$$\Sigma X \xrightarrow{\Sigma g} \Sigma(Y^{(n)}) \xrightarrow{\gamma_n(Y)} (\Sigma Y)^{(n)}.$$

For two maps $f: X \to Y^{(n)}$ and $g: Y \to Z^{(m)}$, one checks that

$$\bar{\Sigma}(g * f) = (\bar{\Sigma} g) * (\bar{\Sigma} f)$$

by using the following commutative diagram:

$$\begin{array}{cccc} \Sigma X & & & \\ \downarrow{\scriptstyle \Sigma f} & & & \\ \Sigma(Y^{(n)}) & \xrightarrow{\gamma_n(Y)} & (\Sigma Y)^{(n)} & \\ \downarrow{\scriptstyle \Sigma(g^{(n)})} & & \downarrow{\scriptstyle (\Sigma g)^{(n)}} & \\ \Sigma(Z^{(m)(n)}) & \xrightarrow{\gamma_n(Z^{(m)})} & (\Sigma Z^{(m)})^{(n)} & \xrightarrow{(\gamma_m(Z))^{(n)}} (\Sigma Z)^{(m)(n)} \\ \downarrow{\scriptstyle \Sigma(\beta^n_m)} & & & \downarrow{\scriptstyle \beta^n_m(\Sigma Z)} \\ \Sigma(Z^{(mn)}) & & \xrightarrow{\gamma_{mn}(Z)} & (\Sigma Z)^{(mn)}. \end{array}$$

The functor $\bar{\Sigma}$ on M induces a functor $\bar{\Sigma}$ on the quotient category $\bar{M}$, because if $f: X \to Y^{(n)}$ is symmetrically equivalent to $g: X \to Y^{(m)}$ then $d^q_n(Y)f$ is homotopic to $d^q_m(Y)g$ for some q. Applying Σ we get a homotopy commutative diagram

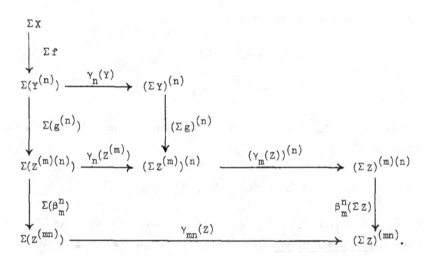

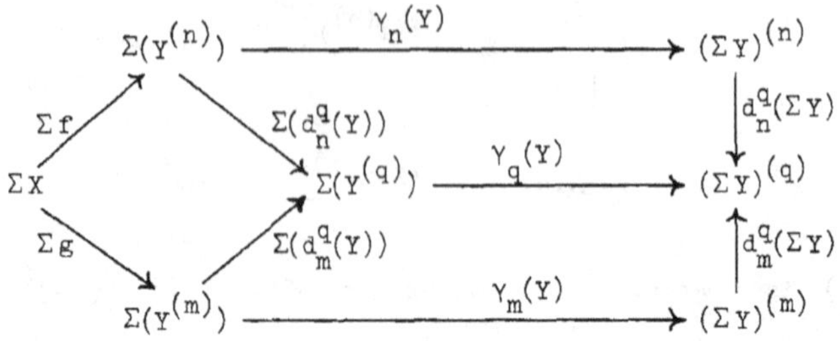

This shows that $\bar{\Sigma} f$ is symmetrically equivalent to $\bar{\Sigma} g$.

§2. <u>The trace homomorphism.</u> Let K be a simplicial set with face maps $d_i\colon K_m \to K_{m-1}$ and degeneracies $s_i\colon K_m \to K_{m+1}$.

(2.1) For a positive integer n, the n-<u>fold</u> (cartesian) <u>product</u> K^n of K is the simplicial set with

$$(K^n)_m = (K_m)^n \qquad (m \geq 0)$$

and with face and degeneracy maps given by

$$d_i(x_1,\ldots,x_n) = (d_i x_1,\ldots,d_i x_n)$$

$$s_i(x_1,\ldots,x_n) = (s_i x_1,\ldots,s_i x_n)$$

where $(x_1,\ldots,x_n) \in (K_m)^n$ and $0 \leq i \leq m$. The symmetric group Σ_n gives a simplicial action on K^n by the rule

$$g(x_1,\ldots,x_n) = (x_{g(1)},\ldots,x_{g(n)})$$

for $g \in \Sigma_n$ and $(x_1,\ldots,x_n) \in (K_m)^n$. Identifying modulo the action of Σ_n, we obtain the n-<u>fold symmetric product</u> $K^{(n)}$ of K, where $(K^{(n)})_m = (K_m)^{(n)}$, $(m \geq 0)$, and

$$d_i[x_1,\ldots,x_n] = [d_i x_1,\ldots,d_i x_n]$$

$$s_i[x_1,\ldots,x_n] = [s_i x_1,\ldots,s_i x_n]$$

for $[x_1,\ldots,x_n] \in (K^{(n)})m$, $0 \leq i \leq m$.

(2.2) Let $|\ |$ denote the geometric realization functor. Then for every positive integer n and any simplicial set K, there are weak homotopy equivalences:

$$|K^n| \to |K|^n \quad \text{and} \quad |K^{(n)}| \to |K|^{(n)}$$

(see [5], p. 162, 6.6).

(2.3) Let $K(X)$ denote the singular complex of a space X. To each m simplex $(\sigma_1,\ldots,\sigma_n)$ in $K(X)^n$, we may associate an m-simplex $\theta(\sigma_1,\ldots,\sigma_m) \in K(X^n)$ by

$$\theta(\sigma_1,\ldots,\sigma_m)(t) = (\sigma_1(t),\ldots,\sigma_n(t))$$

(for t in the standard simplex Δ_m) to give a simplicial map

$$K(X)^n \xrightarrow{\theta} K(X^n).$$

The symmetric group Σ_n acts upon both $K(X)^n$ and $K(X^n)$, and θ is equivariant. Hence a simplicial map $\bar{\theta}$ is induced so that the diagram commutes

$$\begin{array}{ccc} K(X)^n & \xrightarrow{\quad\theta\quad} & K(X^n) \\ \downarrow & & \downarrow K(p) \\ K(X)^{(n)} & \xrightarrow{\quad\bar{\theta}\quad} & K(X^{(n)}) \end{array}$$

where $p\colon X^n \to X^{(n)}$ is the projection.

(2.4) For an arbitrary simplicial set K, let $C(K)$ denote the chain complex of K; that is, $C_m(K)$ is the free abelian group on the m-simplexes of K, $m \geq 0$, with boundary given by

$$\partial(x) = \sum_{i=0}^{m} (-1)^i d_i x$$

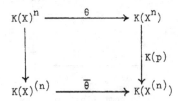

for $x \in C_m(K)$. The homology of $C(K)$ is, by definition, the homology $H(K)$ of K. Then $H(K(X))$ is the singular homology $H(X)$ of a space X. The map $\bar{\theta}$ induces a natural isomorphism

$$\bar{\theta}_*: \quad H(K(X)^{(n)}) \to H(X^{(n)}).$$

(See [5], p. 406, 7.6.25) for any space X and integer $n > 0$. Henceforth, we will identify $H(X^{(n)})$ with $H(K(X)^{(n)})$.

Given a space X and integer $n > 0$ consider the chain map

$$\bar{\mu}^n(X): \quad C(K(X)^{(n)}) \to C(K(X))$$

given on generators by

$$\bar{\mu}^n_m[x_1,\dots,x_n] = x_1 + \cdots + x_n \tag{2.5}$$

where $[x_1,\dots,x_n] \in (K(X)^{(n)})_m$, $m \geq 0$. The induced homology homomorphism $\mu^n_m(X)$: $H_m(X^{(n)}) \to H_m(X)$ will be called the trace homomorphism.

The trace satisfies the naturality condition: that is, for any map $g: X \to Z$ we get commutativity in the diagram

$$\begin{array}{ccc} H(X^{(n)}) & \xrightarrow{g_*^{(n)}} & H(Z^{(n)}) \\ \downarrow{\scriptstyle \mu^n(X)} & & \downarrow{\scriptstyle \mu^n(Z)} \\ H(X) & \xrightarrow{g_*} & H(Z) \end{array} \tag{2.6}$$

For any space Y, the trace is compatible with the d maps of (1.6) in this sense: Let $n|q$. Then

$$\begin{array}{ccc} H(Y^{(n)}) & \xrightarrow{\mu^n(Y)} & H(Y) \\ \downarrow{\scriptstyle (d^q_n(Y))_*} & & \downarrow{\scriptstyle (q/n)} \\ H(Y^{(q)}) & \xrightarrow{\mu^q(Y)} & H(Y) \end{array} \tag{2.7}$$

commutes, where the homomorphism on the right is multiplication by the integer q/n.

The trace is compatible with the map $\beta^n_m(X)$: $(X^{(m)})^{(n)} \to X^{(m+n)}$ as indicated in the following commutative diagram:

$$(2.8)\qquad \begin{array}{ccc} H((X^{(m)})^{(n)}) & \xrightarrow{\beta^n_m(X)_*} & H(X^{(m\cdot n)}) \\ \Big\downarrow{\scriptstyle \mu^n(X^{(m)})} & & \Big\downarrow{\scriptstyle \mu^{m\cdot n}(X)} \\ H(X^{(m)}) & \xrightarrow{\mu^m(X)} & H(X). \end{array}$$

The trace is compatible with the map $\alpha^m_n(X)\colon\ X^{(n)} \times X^{(m)} \to X^{(n+m)}$ in this way: Consider the diagram, where p_1 and p_2 are coordinate projections.

$$\begin{array}{ccccc} H(X^{(n)} \times X^{(m)}) & & \xrightarrow{p_{1_*}} & & H(X^{(n)}) \\ & \searrow^{\alpha^m_n(X)_*} & & & \\ \Big\downarrow{\scriptstyle p_{2_*}} & & H(X^{(n+m)}) & & \Big\downarrow{\scriptstyle \mu^n(X)} \\ & & & \searrow^{\mu^{n+m}(X)} & \\ H(X^{(m)}) & & \xrightarrow[\mu^m(X)]{} & & H(X) \end{array}$$

Then for $x \in H(X^{(n)} \times X^{(m)})$,

$$(2.9)\qquad \mu^{n+m}(X)\alpha^m_n(X) = \mu^n(X)p_{1_*} + \mu^m(X)p_{2_*}.$$

The proofs of (2.8) and (2.9) are obtained by verifying corresponding diagrams of chain maps on the singular chain complexes.

(2.10) There is also a compatibility condition relating the trace with the homology suspension isomorphism, given in (2.11) below. For a simplicial set K, we may define the <u>suspension</u>, ΣK, as follows:

$$(\Sigma K)_0 = \{a^0, b^0\}, \quad \text{a two element set.}$$

For $j > 0$, $(\Sigma K)_j = (K_{j-1})^+ \cup \{a^j, b^j\}$, where $(K_{j-1})^+$ denotes a set in one-one correspondence with K_{j-1} by a map $\sigma \longmapsto \sigma^+$, $(\sigma \in K_{j-1})$, and $\{a^j, b^j\}$ is a two element set. The face and degeneracy maps are given by the rules (where $\sigma \in K_k$):

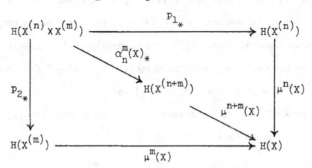

$$d_k(\sigma^+) = b^{k-1}, \quad \text{and} \quad d_i(\sigma^+) = (d_i\sigma)^+ \quad \text{if} \quad 0 \leq i < k,$$

$$s_k(\sigma^+) = b^{k+1}, \quad \text{and} \quad s_i(\sigma^+) = (s_i\sigma)^+ \quad \text{if} \quad 0 \leq i < k,$$

and

$$d_i(a^k) = a^{k-1}, \quad d_i(b^k) = b^{k-1},$$

$$s_i(a^k) = a^{k+1}, \quad s_i(b^k) = b^{k+1} \quad \text{for all} \quad 0 \leq i \leq k.$$

For a space Z, $K(Z)$ will denote its singular complex, $S(Z) = CK(Z)$ (see (2.4)) its singular chain complex, and $S(Z)^+ = C\Sigma K(Z)$ the chain complex of its suspended singular complex. Then $S(\Sigma Z)$ is naturally chain equivalent to $S(Z)^+$ and we obtain the homology suspension isomorphism $H_{q+1}(Z) \simeq H_q(Z)$ for all q.

For a space X, a generator $[\sigma_1,\dots,\sigma_n]$ of $S_q(X^{(n)})$, where each σ_i is a singular q-simplex of X, will correspond to the generator $[\sigma_1,\dots,\sigma_n]^+$ of $S_{q+1}(X^{(n)})^+$. Then the map (see (1.17))

$$\gamma_n(X)\colon \ \Sigma(X^{(n)}) \to (\Sigma X)^{(n)}$$

induces a chain map $S\gamma_n(X)$, which, when regarded as a chain map defined on a suspended chain complex, becomes

$$S\gamma_n(X)\colon \ S(X^{(n)})^+ \to C((\Sigma K(X))^{(n)})$$

and is given on generators by

$$S\gamma_n(X)[\sigma_1,\dots,\sigma_n]^+ = [\sigma_1^+,\dots,\sigma_n^+]$$

for $\sigma_1,\dots,\sigma_n$ q-simplexes of X.

The chain map $\bar{\mu}^n(X)\colon S(X^{(n)}) \to S(X)$ (see 2.5) induces a chain map $\bar{\mu}^n(X)^+\colon S(X^{(n)})^+ \to S(X)^+$ which is given by $\bar{\mu}^n(X)^+[\sigma_1,\dots,\sigma_n]^+ = \sigma_1^+ + \dots + \sigma_n^+$ for singular q-simplexes $\sigma_1,\dots,\sigma_n$ of X.

We obtain a commutative diagram of chain maps:

$$\begin{array}{ccc}
S_{q+1}(\Sigma(X^{(n)})) & & \\
\| & & \\
S_{q+1}(X^{(n)})^{+} & \xrightarrow{S\gamma_n(X)} & S_{q+1}((\Sigma X)^{(n)}) \\
\downarrow{\scriptstyle \overline{\mu}^{n}(X)} & & \downarrow{\scriptstyle \overline{\mu}^{n}(\Sigma X)} \\
S_{q+1}(X)^{+} & = & S_{q+1}(\Sigma X)
\end{array}$$

for all spaces X and all q and $n > 0$. Passing to homology we obtain the compatibility condition of trace and suspension:

(2.11)
$$\begin{array}{ccc}
H_q(X^{(n)}) & \simeq & H_{q+1}(\Sigma(X^{(n)})) \\
 & & \downarrow{\scriptstyle \gamma_n(X)_*} \\
\downarrow{\scriptstyle \mu^{n}(X)} & & H_{q+1}((\Sigma X)^{(n)}) \\
 & & \downarrow{\scriptstyle \mu^{n}(\Sigma X)} \\
H_q(X) & \simeq & H_{q+1}(\Sigma X) .
\end{array}$$

Given $f\colon X \to Y^{(n)}$ and $\overline{f}\colon X \to Y^{(\overline{n})}$ which are symmetrically equivalent, the two homomorphisms $\mu^{n}f_*$ and $\mu^{\overline{n}}\overline{f}_*$ from $H(X)$ to $H(Y)$ need not agree. But they do agree "up to integer multiples" as follows:

(2.12) LEMMA. _For some positive integer q which n and $\overline{n}$ both divide,_

$$(q/\overline{n})\mu^{\overline{n}}(Y)\overline{f}_* = (q/n)\mu^{n}(Y)f_*.$$

PROOF. Since f and $\overline{f}$ are symmetrically equivalent, $d^{q}_{n}(Y)f$ is homotopic to $d^{q}_{\overline{n}}(Y)\overline{f}$ for some q which is a multiple of n and $\overline{n}$. Using (2.7), we get commutativity in the following diagram:

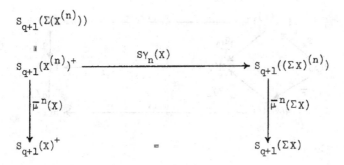

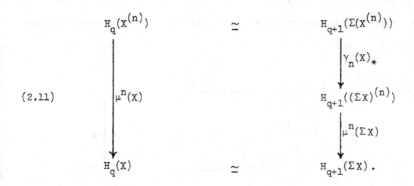

$$\begin{array}{ccccc}
 & & H(Y^{(n)}) & \xrightarrow{\mu^{n}(Y)} & H(Y) \\
 & \overset{f_*}{\nearrow} & & \overset{d^q_n(Y)_*}{\searrow} & \downarrow q/n \\
H(X) & & & & H(Y^{(q)}) \longrightarrow H(Y) \\
 & \underset{\bar{f}_*}{\searrow} & & \underset{d^q_{\bar{n}}(Y)_*}{\nearrow} & \downarrow q/\bar{n} \\
 & & H(Y^{(\bar{n})}) & \xrightarrow{\mu^{\bar{n}}(Y)} & H(Y)
\end{array}$$

and the lemma is proven. ||

(2.13) The correspondence which associates to each space X its homology group and to each map $X \overset{f}{\to} Y^{(n)}$ the homomorphism $H(X) \xrightarrow{\mu^n(Y)f_*} H(Y)$ is a functor from M (see (1.12)) to graded abelian groups.

That the correspondence preserves composites may be checked using the commutative diagram (see (2.8) and (2.6)):

$$\begin{array}{ccccc}
H(X) & \xrightarrow{f_*} & H(Y^{(n)}) & \xrightarrow{\mu^n(Y)} & H(Y) \\
 & & \Big\downarrow g^{(n)}_* & & \Big\downarrow g_* \\
 & & H((Z^{(m)})^{(n)}) & \xrightarrow{\mu^n(Z^{(m)})} & H(Z^{(m)}) \\
 & & \Big\downarrow \beta^n_m(Z)_* & & \Big\downarrow \mu^m(Z) \\
 & & H(Z^{(mn)}) & \xrightarrow{\mu^{mn}(Z)} & H(Z) .
\end{array} \tag{2.14}$$

Homotopic multimaps correspond to the same homology homomorphisms, but symmetrically equivalent multimaps do not, in general.

We observe that there is a similarity between the multivalued function theory of Jerrard (see [3]) and the theory presented here. Our approach is less general, but allows one to use the machinery of algebraic topology more readily.

§3. Degree of a multimap. Let $f\colon S^q \to (S^q)^{(n)}$ be a multimap on the q-sphere for some $q > 0$ and $n > 0$. Then $\mu^n(S^q)f_*\colon H_q(S^q) \to H_q(S^q)$ is a homomorphism of infinite cyclic groups and may be characterized by an integer a such that

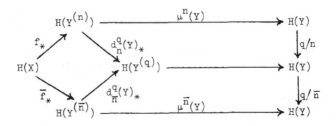

$\mu^n(S^q)f_*(Z) = aZ$ for all $Z \in H_q(S^q)$.

(3.1) The <u>degree</u> <u>of</u> f, deg f, is the rational number a/n.

Suppose $\bar{f}\colon S^q \to (S^q)^{(\bar{n})}$ is symmetrically equivalent to $f\colon S^q \to (S^q)^{(n)}$ and $\mu^{\bar{n}}(S^q)\bar{f}_*(z) = \bar{a}Z$ for all $Z \in H_q(S^q)$. Then using (2.12) there is a common multiple q of n and $\bar{n}$ for which

$$(q/\bar{n})\bar{a}z = (q/n)az$$

for all $z \in H_q(S^q)$. Hence, $\bar{a}/\bar{n} = a/n$, and we may define <u>degree</u> <u>of</u> [f] by deg[f] = deg f.

We now establish the properties of degree listed in the introduction, the first of which is obvious.

(3.2) PROPOSITION. <u>For a map</u> $f\colon S^q \to S^q$ <u>into the</u> 1-<u>fold symmetric product</u>, deg f <u>is the integer value agreeing with the usual degree</u>. <u>In particular, degree of the identity is</u> 1.

(3.3) PROPOSITION. <u>For multimaps</u> g <u>and</u> f <u>on</u> S^q, $\deg g * f = (\deg g)(\deg f)$.

PROOF. Let $f\colon S^q \to (S^q)^{(n)}$ and $g\colon S^q \to (S^q)^{(m)}$. Referring to diagram (2.14) with $X = Y = Z = S^q$ and $H = H_q$, we have for every $z \in H_q(S^q)$,

$$\mu^{mn}(S^q)\beta^n_m(S^q)_*g^{(n)}{}_*f_*(z) = \mu^m(S^q)g_*\mu^n(S^q)f_*(z)$$

$$m \cdot n \deg(g * f)z = m(\deg g)n(\deg f)z.$$

Hence

$$\deg(g * f) = (\deg g)(\deg f). \quad ||$$

(3.4) APPLICATION. For an integer $a \neq 0$, the $|a|$-fold covering $u_a\colon S^1 \to S^1$ has degree = a, by (3.2). Therefore the multimap v_a of (1.16) with the property that $[u_a] * [v_a] = [1_{S^1}]$ has degree 1/a, since a $\deg v_a = \deg u_a \cdot \deg v_a = \deg(u_a * v_a) = \deg[u_a * v_a] = \deg[u_a] * [v_a] = [1]$. For integers a and b, with $b \neq 0$, it follows that $\deg u_a * v_b = (a/b)$.

(3.5) PROPOSITION. *For any multimap* f *on* S^q, $\deg f = \deg \bar{\Sigma} f$, *where* $\bar{\Sigma}$ *is the suspension functor of* (1.19).

PROOF. Consider the following diagram where the horizontal arrows are homology suspension isomorphisms.

$$\begin{array}{ccc}
H_{q+1}(\Sigma S^q) & \xrightarrow{\cong} & H_q(S^q) \\
\downarrow{\scriptstyle (\Sigma f)_*} & & \downarrow{\scriptstyle f_*} \\
H_{q+1}(\Sigma((S^q)^{(n)})) & \xrightarrow{\cong} & H_q((S^q)^{(n)}) \\
\downarrow{\scriptstyle \gamma_n(S^q)_*} & & \\
H_{q+1}((\Sigma S^q)^{(n)}) & & \downarrow{\scriptstyle \mu^n(S^q)} \\
\downarrow{\scriptstyle \mu^n(\Sigma S^q)} & & \\
H_{q+1}(\Sigma S^q) & \xrightarrow{\cong} & H_q(S^q)
\end{array}$$

By naturality of the suspension isomorphism and by (2.11), the diagram commutes; hence (3.5) follows. ||

(3.6) APPLICATION. Referring to the notation of (1.15) and (3.4), the multimap $u_a * v_b$ on S^1 has degree a/b where a and b are integers and $b \neq 0$. By successive suspensions a multimap of degree a/b can be obtained on S^q for any $q > 0$. This proves the next result.

(3.7) PROPOSITION. *The correspondence from multimaps on* S^q *to the rational numbers given by* $f \mapsto \deg f$ *is an epimorphism for all* $q > 0$. ||

(3.8) PROPOSITION. *If* f *and* g *are multimaps on* S^q *of multiplicities* n *and* m, *respectively, then the degree of the join (see* (1.16)) *is given by*

$$\deg(f \vee g) = \frac{n(\deg f) + m(\deg g)}{n + m} .$$

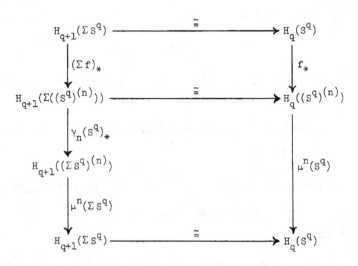

PROOF. Let X denote S^q. Let p_1 and p_2 be the first and second projections of the cartesian product $X^{(n)} \times X^{(m)}$ onto its coordinate spaces. Using (2.9), we obtain

$$\mu^{(n)}(X)p_{1*} + \mu^{(m)}(X)p_{2*} = \mu^{(n+m)}(X)\alpha_n^m(X)_*,$$

which are homomorphisms from $H_q(X^{(n)} \times X^{(m)})$ to $H_q(X)$.

Now for $z \in H_q(X)$, $(n+m)\deg(f \vee g)(z) = \mu^{(n+m)}(X) \cdot (f \vee g)_*(z) = \mu^{(n+m)}(X)\alpha_n^m(X)_*(f \times g)_*(z) = (\mu^{(n)}(X)p_{1*} + \mu^{(m)}(X)p_{2*})(f \times g)_*(z) = \mu^{(n)}(X)f_*(z) + \mu^{(m)}(X)g_*(z) = n \deg f + m \deg g$. ||

(3.9) APPLICATION. The weighted average formula of (3.8) has an obvious extension to the join of multimaps $f_i\colon S^q \to (S^q)^{(n_i)}$, $i = 1,\dots,k$, $k \geq 2$. We have:

$$\left(\sum_{i=1}^{k} n_i\right)\deg(f_1 \vee \cdots \vee f_k) = \sum_{i=1}^{k} n_i \deg f_i .$$

The map $A_n\colon S^q \to (S^q)^{(n)}$ given by $A_n(x) = [-x, -x,\dots,-x]$ for $x \in S^q$ is called the <u>antipodal</u> <u>multimap</u> of multiplicity n. Then $A_1\colon S^q \to S^q$ has degree $(-1)^{q+1}$ (see [2], p. 66, 4.3) and $A_n = A_1 \vee \cdots \vee A_1$ (n factors). The averaging formula gives $\deg A_n = (-1)^{q+1}$ for all n. ||

A multimap f on S^q has a <u>fixed point</u> if x is a coordinate of $f(x)$ for some $x \in S^q$.

(3.10) PROPOSITION. <u>If a multimap</u> f <u>on</u> S^q <u>has no fixed point, then</u> $\deg f = (-1)^{q+1}$.

PROOF. We will construct a homotopy between f and A_n. Let $E = R^{q+1}$ and let $0 \in E$ be the origin. Scalar multiplication $R \times E \to E$ induces a map $R \times E^n \to E^n$ given by $(t, (y_1,\dots,y_n)) \longmapsto (ty_1,\dots,ty_n)$ which is equivariant with respect to the symmetric group action. Hence the map $R \times E^{(n)} \to E^{(n)}$ where $(t, [y_1,\dots,y_n]) \longmapsto [ty_1,\dots,ty_n]$ is well defined and continuous.

Similarly, the difference map $E^n \times E \to E^n$ given by $((y_1,\dots,y_n), x) \longmapsto (y_1 - x,\dots,y_n - x)$ induces a well-defined and continuous map $d\colon E^{(n)} \times E \to {}^{(n)}$ where

$$d([y_1,\ldots,y_n], x) = [y_1 - x,\ldots,y_n - x].$$

Note that $d(y,0) = y$ and $d([0,\ldots,0], x) = A_n(x)$ for $y \in E^{(n)}$, $x \in E$.

Since f has no fixed points, it follows that for every $x \in S^q$, $d((1-t)f(x), tx)$ has no coordinate equal to zero, $0 \leq t \leq 1$. Hence, for $x \in S^q$ and $t \in I$, $d((1-t)f(x), tx) \in (E-0)^{(n)}$. Define $F\colon S^q \times I \to (S^q)^{(n)}$ by $F(x,t) = r^{(n)}d((1-t)f(x), tx)$ where $r\colon E - 0 \to S^q$ is the retraction given by $r(y) = \frac{y}{\|y\|}$. Then F is a homotopy from f to the antipodal map A_n. Hence, $\deg f = (-1)^{q+1}$. ||

A multimap f on S^q has an <u>antipodal point</u> if $-x$ is a coordinate of $f(x)$ for some $x \in S^q$.

(3.11) COROLLARY. *If a multimap f on S^q has no antipodal point, then* $\deg f = 1$.

PROOF. If $f\colon S^q \to (S^q)^{(n)}$ has no antipodal point, then the scalar multiple $-f$ has no fixed point. But $-f = A_1 * f$. Hence

$$(-1)^{q+1} = \deg(-f) = \deg A_1 * f = \deg A_1 \deg f = (-1)^{q+1}\deg f.$$

(3.12) COROLLARY. *If q is even, every multimap on S^q has a fixed point or an antipodal point.*

The results (3.10) and (3.12) may also be obtained by using the Lefshetz number associated with the induced homology homomorphism $\mu^n(S^q)f_*$ of the multimap $f\colon S^q \to (S^q)^{(n)}$ (see [4]).

REFERENCES

1. R. Brown, Elements of Modern Topology, McGraw-Hill, 1968.

2. A. Dold, Lectures on Algebraic Topology, Springer, 1972.

3. R. P. Jerrard, Homology with multiple-valued functions applied to fixed points, Trans. Amer. Math. Soc. vol. 213 (1975), 407-427.

4. C. Maxwell, Fixed points of symmetric product mappings, Proc. Amer. Math. Soc. vol. 8 (1957), 808-815.

5. E. Spanier, Infinite symmetric products, function spaces and duality, Annals of Math. vol. 69 (1959), 142-166.

STRATIFIED GENERAL POSITION

by Clint McCrory
Brown University

In the thirties, Lefschetz' intersection theory for manifolds was generalized to spaces with singularities by the introduction of cohomology and cup product. Recently Goresky and MacPherson have defined a new homology theory for stratified spaces, with an intersection pairing which extends the cup product pairing [GM].

In this note, I prove a general position theorem for polyhedra which can be used to define the Goresky-MacPherson pairing, in the same way that the classical general position theorem gives a way to intersect homology classes in a polyhedral manifold. This theorem was proved in my thesis [M1], refining earlier work of Akin [A]. For the necessary background in piecewise-linear topology, I refer the reader to Rourke and Sanderson's book [RS].

§1. Stratified polyhedra.

A filtered polyhedron $(X) = (X_n, X_{n-1}, \dots, X_0)$ is a polyhedron X together with a family of closed subpolyhedra $X = X_n \supset X_{n-1} \supset \dots \supset X_0$. A map $f:(X) \to (Y)$ of filtered polyhedra is a piecewise-linear map $f: X \to Y$ such that $f(X_i) \subset Y_i$ for all i.

A stratified polyhedron is a filtered polyhedron $(X) = (X_n, \dots, X_0)$ such that for each point $x \in X_i \setminus X_{i-1}$ there exists a compact filtered polyhedron $(L) = (L_n, \dots, L_{i+1})$ and a PL homeomorphism of filtered polyhedra from a neighborhood of x in (X) to

$$D^i \times c(L_n, \dots, L_{i+1}, \emptyset) = (D^i \times cL_n, \dots, D^i \times cL_{i+1}, D^i \times c) .$$

Here D^i is an i-disc and cY denotes the cone on Y. The subpolyhedron $\chi = X_i \setminus X_{i-1}$ is thus a (possibly empty) PL i-manifold, the ith stratum of (X). The filtered polyhedron (L) is the link of χ_i at the point x. Note that $\dim L_k = k - i - 1$. For further discussion of PL stratifications, see [S] and [M2].

For example, the filtration of X by the skeletons of a PL triangulation is a stratification. At the other extreme, every polyhedron X has an intrinsic stratification $(I^n(X),\ldots,I^0(X))$, characterized by the property that if $(X_n,\ldots,X_0)$ is any stratification of X, then $I^i(X) \subset X_i$ for all i (cf. [A]). Every Whitney stratified space is homeomorphic with a stratified polyhedron (cf. [G]).

§2. General position.

The subpolyhedra A and B of X are in general position in the stratification (X) if

$$\dim(A \cap B \cap \chi_i) \leq \dim(A \cap \chi_i) + \dim(B \cap \chi_i) - i$$

for all i. In other words, $A \cap \chi_i$ and $B \cap \chi_i$ are in general position in the i-manifold χ_i for all i.

A PL isotopy H of (X) is a PL isotopy H of X such that $H_t(X_i) \subset X_i$ for each i.

Theorem. Let (X) be a stratified polyhedron, and let A, B, C, be closed subpolyhedra with $B \supset C$. Given $\epsilon > 0$ there exists a PL isotopy H of (X) such that

1) $|H_t(x) - x| < \epsilon$ for all x and t,
2) $H_t(x) = x$ for all $x \in C$ and all t,
3) A and $H_1(B \setminus C)$ are in general position in (X).

This theorem was proved, without condition (2), in my thesis [M1, p. 98]. It was proved first for the intrinsic stratification, without condition (1), in Akin's thesis [A, Th. 6, p. 471].

§3. Proof of the theorem.

The proof uses the simplicial technique of Zeeman [Z, chapter 6]. The isotopy H will be constructed as the composition of a finite sequence of "shifts".

Let (X), A, B, C, and ϵ be as above. If j and k are integers with $0 \leq j \leq k \leq n = \dim(X)$, an ϵ (j,k)-<u>shift</u> F <u>of</u> B <u>with respect</u> to A <u>in</u> (X) <u>keeping</u> C <u>fixed</u> is an isotopy of (X) defined as follows. Let K be a triangulation of X so that the subpolyhedra A, B, C, and X_i for all i are subcomplexes. (To simplify notation, I will assume that $|K| = X$, and refer to the simplexes of K as subsets of X.) Let K' denote the barycentric subdivision of K modulo the $(j-1)$-skeleton, obtained by starring each simplex Δ of dimension $\geq j$ at its barycenter $b(\Delta)$, in order of decreasing dimension. Choose K so that the diameter of $\mathrm{star}(b(\Delta),K')$ is less than ϵ. For each j-simplex Δ of K, the isotopy F_t will map $\mathrm{star}(b(\Delta),K')$ to itself, keeping the frontier fixed. Since X is the union of $\mathrm{star}(b(\Delta),K')$ over all j-simplexes Δ, and the intersection of two such stars is contained in the intersection of their frontiers, F_t can be defined by specifying its restriction to the star of each j-simplex Δ. Let F_t be the identity on $\mathrm{star}(b(\Delta),K')$ unless Δ is contained in $(B \setminus C) \cap X_k$.

If Δ is a j-simplex of $(B \setminus C) \cap X_k$, then the restriction of F to $\mathrm{star}(b(\Delta),K')$ is a modification of Zeeman's "local shift". Let J_k and K_k be the subcomplexes of K with $|J_k| = B \cap X_k$ and $|K_k| = X_k$. Let $P = \mathrm{link}(b(\Delta),J_k')$ and $Q = \mathrm{link}(b(\Delta),K_k')$. Since $\Delta \subset X_k$, the link Q is a $(k-1)$-sphere. If $b = b(\Delta)$ and $\dot{\Delta}$ is the boundary of Δ, then $\mathrm{star}(b(\Delta),J_k')$ is the join $b\dot{\Delta}P$ and $\mathrm{star}(b(\Delta),K_k')$ is the join $b\dot{\Delta}Q$. The isotopy $F_t|\mathrm{star}(b(\Delta),K_k')$ is constructed by "shifting" the embedding $b\dot{\Delta}P \to b\dot{\Delta}Q$ exactly as in Zeeman's proof [Z, ch. 6, p. 11-12]. ("The basic idea is to reduce the intersection dimension of two cones in Euclidean space by moving their vertices slightly apart." [Z, ch. 6, p. 5]. Notice that A enters into the construction of the shift in choosing the triangulation K so that A is a subcomplex.)

Now let (L) be the link of the stratum X_k at $b(\Delta)$. By the definition of a stratification, there is a PL homeomorphism $h : \mathrm{star}(b(\Delta),K') \to$

$(\mathrm{star}(b(\Delta),K_k'))*L$ such that $h(\mathrm{star}(b(\Delta),K_k')) = \mathrm{star}(b(\Delta),K_k')$ and $h(\mathrm{star}(b(\Delta),K_i')) = (\mathrm{star}(b(\Delta),K_k'))*L_i$ for each $i > k$. Let $F_1|\mathrm{star}(b(\Delta),K_k') = h^{-1}T_\Delta h$, where T_Δ is the join of $F_1|\mathrm{star}(b(\Delta),K_k')$ with the identity map of L. The isotopy $F_t|\mathrm{star}(b(\Delta),K_k')$ extends to an isotopy $F_t|\mathrm{star}(b(\Delta),K')$ by lemma 1 below. This completes the definition of the (j,k)-shift F.

Lemma 1. Let P and Q be compact polyhedra. Let F be a PL isotopy of P and let G be a PL isotopy of Q. Then there is a PL isotopy H of the join $P*Q$ such that $H_1 = F_1 * G_1$ and H restricts to F on P and to G on Q.

Proof. It is clear from the definition of simplicial join that

$$(P*Q) \times I = [(P \times I)*(Q \times \{1\})] \cup [(P \times \{0\})*(Q \times I)],$$

where the union is along the common subpolyhedron $(P \times \{0\})*(Q \times \{1\})$. Let $H|(P \times I)*(Q \times \{1\})$ be the join of F with G_1, and let $H|(P \times \{0\})*(Q \times I)$ be the join of the identity with G.

Lemma 2. Let F be an ϵ (j,k)-shift of B with respect to A in (X) keeping C fixed.

1) $|F_t(x) - x| < \epsilon$ for all x and t,

2) $F_t(x) = x$ for all $x \in C \cup X_{k-1}$ and all t,

3a) If $(B \setminus C) \cap X_k$ is in general position with $A \cap X_k$ in X_k, then so is $F_1(B \setminus C) \cap X_k$.

3b) If $\dim(A \cap (B \setminus C) \cap X_k) = j > \dim(A \cap X_k) + \dim((B \setminus C) \cap X_k) - k$, then $\dim(A \cap F_1(B \setminus C) \cap X_k) = j - 1$.

Proof. Properties (1) and (2) are clear from the construction of F. The proof of (3a) and (3b) is exactly the same as the proof of the corresponding lemma of Zeeman [Z, ch. 6, p. 12].

Now I shall construct the isotopy H. Let (X), A, B, C, and ϵ be as above. First I construct a sequence $G^1, \ldots, G^n$ of isotopies of (X) such that for each i

1) $|G_t^i(x) - x| < \varepsilon/n$ for all x and t,

2) $G_t^i(x) = x$ for all $x \in C \cup X_{i-1}$ and all t,

3) $\dim(A \cap G_1^i(B^{i-1} \setminus C) \cap \chi_i) \leq \dim(A \cap \chi_i) + \dim(B^{i-1} \setminus C) \cap \chi_i) - i$,

where $B^{i-1} = G_1^{i-1} G_1^{i-2} \ldots G_1^1(B)$.

The product of these isotopies will then be an isotopy H with properties (1) (2) (3) of the theorem.

Suppose that G^i has been defined for $i = 1, \ldots, k-1$. Let $s = \dim(A \cap (B^{k-1} \quad C) \cap \chi_k)$ and $t = \dim(A \cap \chi_k) + \dim(B^{k-1} \cap \chi_k) - k$. If $s \leq t$ let G^k be the identity isotopy. If $s > t$, let G^k be the product of isotopies $F^{t+1,k} F^{t+2,k} \ldots F^{s,k}$, where $F^{j,k}$ is an $\varepsilon/(s-t)n$ shift of $F^{j+1,k} F^{j+2,k} \ldots F^{s,k}(B^{k-1})$ with respect to A in (X) keeping C fixed. Then (1) (2) (3) for G^k follow from the corresponding statements in lemma 2. This completes the proof of the theorem.

BIBLIOGRAPHY

[A] E. Akin, Manifold phenomena in the theory of polyhedra, Trans. Amer. Math. Soc. 143 (1969), 413-473.

[G] M. Goresky, Triangulation of stratified sets, to appear.

[GM] M. Goresky; R. MacPherson, La dualité de Poincaré pour les espaces singuliers, C. R. Acad. Sci. Paris, série A, 284 (1977), 1549-1551.

[M1] C. McCrory, Poincaré duality in spaces with singularities, Ph.D. thesis, Brandeis University, May, 1972.

[M2] C. McCrory, Cone complexes and PL transversality, Trans. Amer. Math. Soc. 207 (1975), 269-291.

[RS] C.P. Bourke; B.J. Sanderson, Introduction to Piecewise-Linear Topology, Springer-Verlag, New York, 1972.

[S] D. Stone, Stratified polyhedra, Lecture Notes in Math., Vol. 252, Springer-Verlag, New York, 1972.

[Z] E.C. Zeeman, Seminar on combinatorial topology, I.H.E.S. Paris and the University of Warwick at Conventry, 1963-1966.

CLIFFORD BUNDLES, SPINOR BUNDLES AND DIRAC OPERATORS

by M.L. Michelsohn

University of California, Berkeley

§1 Introduction.

This is a report on joint work currently in progress with Blaine Lawson. We will begin with an exposition of basic concepts and some known results. We will then discuss some new applications to geometry and topology. In §3 we outline some immersion and embedding results; in §4 we make applications to minimal submanifolds and in §5 we develop a Clifford cohomology theory for Kähler manifolds.

Recall that for any differentiable manifold M one can consider the bundle $\wedge^*(M)$ of exterior forms with the exterior derivative d. This bundle of algebras is fundamental to the study of the topological structure of M. For example, Sullivan [9] has proved that the sections of $\wedge^*(M)$, considered as a graded differential algebra, determine the real homotopy type of M. Now if M is a riemannian manifold, then via the riemannian metric one can define in a natural way another bundle of algebras $Cl(M)$ called the Clifford bundle of M (with an associated Dirac operator D). This bundle carries more information than $\wedge^*(M)$ and plays an important role in the study of the riemannian structure of M. In fact, $\wedge^*(M)$ and $Cl(M)$ are canonically isomorphic as vector bundles, but not as bundles of algebras. Multiplication in $Cl(M)$ carries a natural filtration for which $\wedge^*(M)$ is the bundle of associated graded algebras. It is our program to investigate the structure of $Cl(M)$ and its applications to problems in riemannian geometry and topology.

§2 Basic concepts.

We begin by recalling the notion of a Clifford algebra. Let V be a vector space with a quadratic form Q. Consider the tensor algebra $T(V) = \sum_{r=0}^{\infty} \otimes^r V$

and let I denote the ideal in $T(V)$ generated by elements of the form $v \otimes v + Q(v)$ for $v \in V$. Then the quotient

$$Cl_Q(V) = T(V)/I$$

is defined to be the <u>Clifford algebra of</u> V <u>with quadratic form</u> Q.

<u>Example 2.1</u>. Suppose $V = R^n$ and $Q(v) = \langle v,v \rangle$ where $\langle \cdot , \cdot \rangle$ is the standard inner product on R^n. Let $e_1, \ldots, e_n$ be any orthonormal basis for R^n. Then $Cl_Q(V)$ is generated as an algebra by $e_1, \ldots, e_n$ subject to the relations: $e_i^2 = -1$ and $e_i e_j + e_j e_i = 0$ for $i \neq j$. We denote this algebra by Cl_n.

There is a natural inclusion $R^n \subset Cl_n$ and the unit vectors in R^n thereby generate a subgroup of the group of units in Cl_n which is denoted Pin_n. One can easily check that conjugation by elements of Pin_n preserves $R^n \subset Cl_n$. This gives a surjective homomorphism $Pin_n \to O_n$ with kernel Z_2. Restricting to identity components we get a homomorphism $Spin_n \to SO_n$. $Spin_n$ is the universal covering group of SO_n and can be expressed as all products of the form $v_1 \ldots v_{2k}$ in Cl_n where $v_1, \ldots, v_{2k}$ are unit vectors in R^n and $k = 1, 2, 3, \ldots$. It follows that any module for the algebra Cl_n determines a representation of $Spin_n$.

Suppose now that E is a real, n-dimensional vector bundle with inner product over M. Then by performing the above construction fibrewise one can define a bundle of algebras $Cl(E)$, called the Clifford bundle of E. This is the bundle over M whose fibre at a point x is $Cl_Q(E_x)$ where E_x is the fibre of E at x and Q is the quadratic form given by the inner product in E_x.

It will be of interest also to consider bundles S of left modules over the bundle of algebras $Cl(E)$. The two main examples are the following:

<u>Example 2.2</u>. <u>The Clifford bundle itself</u>. Here the multiplication is just Clifford multiplication on the left.

<u>Example 2.3</u>. <u>Bundles of "spinors"</u>: Suppose E is a spin bundle, i.e., suppose

the Stiefel-Whitney classes $w_1(E) = 0$ and $w_2(E) = 0$. Let $P(E)$ denote the principal SO_n-bundle of oriented orthonormal frames in E. Then (see Milnor [7], for example) there exists a principal $Spin_n$ bundle $\widetilde{P(E)}$ and a two-fold covering $\widetilde{P(E)} \to P(E)$ which is fibrewise the double covering $Spin_n \to SO_n$. Let $\Delta : Spin_n \to SO_N$ be a representation determined by an irreducible Cl_n-module as above. (Up to equivalence there are exactly one or two of these depending on n (mod 4). See [2].) Form the associated bundle

$$S_\Delta(E) = \widetilde{P(E)} \times_\Delta \mathbb{R}^N$$

(i.e., the quotient of $\widetilde{P(E)} \times \mathbb{R}^N$ by the relation $(p,v) \sim (pg^{-1}, \Delta(g)v)$ for $g \in Spin_n$). Then one can show the following:

<u>Proposition 2.4</u>. $S_\Delta(E)$ <u>is a bundle of left modules over</u> $Cl(E)$.

We shall be concerned for the most part with the tangent bundle T of a riemannian manifold M. We have then a canonical riemannian connection on T. This induces a connection on $Cl(M) (= Cl(T))$ which is a derivation with respect to Clifford multiplication. Moreover, if M is a spin manifold and $S(M) (= S_\Delta(T))$ is a bundle of spinors on M, there exists a canonical riemannian connection on $S(M)$ which is a derivation with respect to the module multiplication.

Now whenever we have S, a bundle of modules over $Cl(M)$, and a connection ∇ on S, we can define a <u>Dirac operator</u> $D : \Gamma(S) \to \Gamma(S)$ by the formula:

$$D = \sum_{j=1}^{n} e_j \cdot \nabla_{e_j} \tag{2.1}$$

where $e_1, \ldots, e_n$ are local, pointwise orthonormal, tangent vector fields and where "$\cdot$" denotes Clifford module multiplication. Observe that the expression (2.1) is independent of the choice of the local frame field $(e_1, \ldots, e_n)$ and so D is well defined globally. It is an elliptic first order operator. Moreover, if the connection is a derivation for the module multiplication, it is also self-adjoint.

Example 2.5. Let $S = Cl(M)$. Under the canonical isomorphism $Cl(M) \cong \wedge^*(M)$ we have that

$$D \cong d + \delta$$

where $d : \wedge^{p-1}(M) \to \wedge^p(M)$ is exterior differentiation and $\delta : \wedge^p(M) \to \wedge^{p-1}(M)$ is its adjoint, given by the formula $\delta = (-1)^{np+n+1} * d *$. It follows that

$$D^2 \cong \triangle$$

where $\triangle = d\delta + \delta d$ is the so-called Hodge Laplacian. Thus D represents a "square-root" of the Laplace operator.

Now the words first order elliptic operator bring to mind the Atiyah-Singer Index Theorem [3]. However, a naive application of this theorem to D yields little interest since the index of any self-adjoint operator is $\dim \ker D - \dim \operatorname{coker} D = \dim \ker D - \dim \ker D^* = 0$. However, using the structure of Clifford algebras one can decompose $Cl(M)$ (and $S(M)$) canonically into subbundles which are mapped one into the other by D, thereby yielding complexes for which there is a non-trivial index. In all the following examples, M is compact.

Example 2.6. The Euler Characteristic. Cl_n has an even-odd Z_2-grading inherited from the length filtration in $T(R_n)$. This yields a splitting $Cl(M) = Cl^{even}(M) \oplus Cl^{odd}(M)$. Let $D^{even} = D\big|_{Cl^{even}}$ and $D^{odd} = D\big|_{Cl^{odd}(M)}$. Then

$$D^{even} : Cl^{even}(M) \to Cl^{odd}(M)$$

and $(D^{even})^* = D^{odd}$. Under the isomorphism $Cl(M) \cong \wedge^*(M)$, we have $Cl^{even}(M) \cong \wedge^{even}(M)$ and $Cl^{odd}(M) \cong \wedge^{odd}(M)$. Moreover, since D is self-adjoint, $\ker D = \ker D^2 = \ker \triangle$. It follows that $\operatorname{index}(D^{even}) = \dim \ker D^{even} - \dim \operatorname{coker} D^{even} = \dim \ker D^{even} - \dim \ker D^{odd} = \chi(M)$.

Example 2.7. The signature. Suppose M is oriented and let $\omega = e_1 \cdots e_n$ where $(e_1, \ldots, e_n)$ is locally any oriented orthonormal frame field for M. Then ω is the section of $Cl(M)$ corresponding to the oriented volume element on M. If

$n \equiv 0 \pmod 4$, we have $\omega^2 = 1$ and $\omega e_j = -e_j\omega$ for all j. Hence, in these dimensions we obtain a splitting $Cl(M) = Cl^+(M) \oplus Cl^-(M)$ where $Cl^{\pm}(M)$ are the ± 1 eigenbundles for L_w, left multiplication by w. Then left multiplication by any vector field maps $Cl^{\pm}(M)$ to $Cl^{\mp}(M)$. We then set $D^{\pm} = D|_{Cl^{\pm}(M)}$ and note that

$$D^+ : Cl^+(M) \to Cl^-(M)$$

and $(D^+)^* = D^-$. A straightforward argument using Hodge theory then shows that $\text{index}(D^+) = \sigma(M)$, the signature of M.

Example 2.8. The $\hat{A}$ genus Let M be a spin manifold of dimension $n \equiv 0 \pmod 4$ and let $S(M)$ be the bundle of spinors constructed from the irreducible module for Cl_n. (There is only one such in these dimensions cf. [2].) Again we can split $S = S^+ \oplus S^-$ where S^+ and S^- are the $+1$ and -1 eigenbundles under multiplication by ω. As in Example 2.7, the Dirac operator D determines an operator

$$D^+ : S^+ \to S^-$$

and from the Atiyah-Singer Index Theorem one has that $\text{index}(D^+) = \hat{A}(M)$, the so-called $\hat{A}$-genus of M. (See Hirzebruch [4] for example.)

§3 Applications to Topology

The elliptic complexes constructed in the last section are useful for studying questions concerning the existence of vector fields and the existence of immersions. The simplest case is the following.

Suppose M admits a nowhere vanishing vector field v and set $e = v/\|v\|$. Let R_e denote right Clifford multiplication by e and observe that $R_e : Cl^{odd}(M) \to Cl^{even}(M)$ and $(R_e)^2 = -1$. We now define a new operator $\tilde{D} = \frac{1}{2}(D + R_e^{-1}DR_e) = \frac{1}{2}(D - R_eDR_e)$. $\tilde{D}$ differs from D by a zero-order operator. Hence, by restriction $\tilde{D}$ defines an operator

$$\tilde{D}^{even} : Cl^{even}(M) \to Cl^{odd}(M)$$

which has the same index as D^{even}, namely $\chi(M)$. The adjoint of this operator is $\tilde{D}^{\text{odd}}$, similarly defined by restriction. Now since $\tilde{D}R_e = R_e\tilde{D}$ we have that $\text{Re}(\ker \tilde{D}) = \ker \tilde{D}$. In particular, $\text{Re} : \ker \tilde{D}^{\text{even}} \xrightarrow{\approx} \ker \tilde{D}^{\text{odd}}$. Hence, $\chi(M) = \text{index}(\tilde{D}^{\text{even}}) = \dim(\ker \tilde{D}^{\text{even}}) - \dim(\ker \tilde{D}^{\text{odd}}) = 0$.

Atiyah has applied a similar argument to the complex in Example 2.7 to prove the following. Let a_{r+1} denote the dimension of an irreducible Cl_r module (cf. [2]).

Theorem 3.1 (Atiyah [1]). *If a 4q-dimensional manifold M admits r everywhere linearly independent vector fields, then*

$$\sigma(M) \equiv 0 \pmod{2a_r}$$

Proof. This argument follows closely the one given above. Here we use the splitting $Cl(M) = Cl^+(M) \oplus Cl^-(M)$ given in Example 2.7. Letting $e_1, \ldots, e_r$ be the given vector fields, which we can, of course, assume to be orthonormal, we now average D over the entire multiplicative group Γ generated by $e_1, \ldots, e_r$. That is, we set

$$\tilde{D} = \frac{1}{|\Gamma|} \sum_{\sigma \in \Gamma} R_\sigma^{-1} D R_\sigma .$$

L_ω and R_σ commute for any σ, hence by restriction we can define operators $\tilde{D}^{\pm} : Cl^{\pm}(M) \to Cl^{\mp}(M)$ which are adjoints of one another. D and $\tilde{D}$ differ by a zero-order operator, hence $\text{index}(\tilde{D}^+) = \text{index}(D^+) = \sigma(M)$. The operators R_{e_j} preserve the bundles $Cl^{\pm}(M)$. They also commute with $\tilde{D}$ and so they act on $\ker(\tilde{D}^+)$ and $\ker(\tilde{D}^-)$. Thus, $\ker(\tilde{D}^+)$ and $\ker(\tilde{D}^-)$ are Cl_r-modules (in fact, Z_2-graded). This gives $\sigma(M) \equiv 0 \pmod{a_{r+1}}$. The Z_2-grading gives $\sigma(M) \equiv 0 \pmod{2a_r}$.

In the same spirit one can prove the following. Recall that the A-genus of a 4q-manifold is an integral class related to the $\hat{A}$-genus by $A(M) = 2^{4q}\hat{A}(m)$ (cf. [4]).

Theorem 3.2. *Suppose* M *is a compact oriented manifold of dimension* $n = 4q$. *If* M *admits an immersion into the sphere* S^{n+k}, *then*

$$2^k \hat{A}(M) \equiv 0 \pmod{2a_{n+k}}. \tag{3.1}$$

Note. We recall [2] that a_r is given by the table:

r	a_r
1	1
2	2
3	4
4	4
5	8
6	8
7	8
8	8

and $a_{r+8} = 16\, a_r$.

PROOF. Let N denote the normal bundle of the immersion $M \to S^{n+k}$. Then $T \oplus N$ is a trivial $(n+k)$-bundle over M. We introduce on $T \oplus N$ the natural metric and direct sum connection inherited from the sphere. We then define a Dirac operator D on the bundle $Cl(T \oplus N)$ by setting

$$D = \sum_{j=1}^{n} e_j \cdot \nabla_{e_j}$$

where $(e_1, \ldots, e_n)$ is any local tangent frame field for M and where ∇ denotes the connection on $T \oplus N$. This operator is elliptic and self-adjoint. As before, we consider the global tangent volume element $\omega = e_1 \cdots e_n$. Since $n \equiv 0 \pmod 4$, $(L_\omega)^2 = 1$ and $L_\omega D = -DL_\omega$. There is a decomposition $Cl(T \oplus N) \cong Cl^+ \oplus Cl^-$ and $D: Cl^+ \to Cl^-$.

Now since $T \oplus N$ is trivial, it admits $n + k$ everywhere orthonormal sections $\epsilon_1, \ldots, \epsilon_{n+k}$. Proceeding as in the proof of Theorem 3.2 we average D over the group Γ generated by $\epsilon_1, \ldots, \epsilon_{n+k}$. This yields an operator $\tilde{D}^+ : Cl^+ \to Cl^-$ whose index is congruent to zero $\mathrm{mod}(a_{n+k})$. However, $\mathrm{index}(\tilde{D}^+) = \mathrm{index}(D^+)$ and by the Atiyah-Singer Theorem [3], $\mathrm{index}(D^+)$ can be computed to be $2^k A(M)$. q.e.d.

<u>Remark 3.3</u>. <u>If M is embedded with codimension $k \equiv 0 \pmod 4$ then 3.2 can be strengthened by one more factor of 2 · i.e., $2^k A(M) \equiv 0 \pmod{4a_{n+k}}$.</u>

<u>Example 3.4</u>. Let M be CP^{2q}. Then $A(M) = (-1)^q \binom{2q}{q} = 2^{\alpha(q)}(\text{odd})$ where $\alpha(q)$ is the number of 1's in the dyadic expansion of q. It is now not hard to verify the following corollary (note that Theorem 3.2 is a cobordism result).

<u>Corollary 3.5</u>. <u>Suppose M is a compact oriented manifold cobordant to CP^{2q}. Then</u>

1) <u>if</u> $\alpha(q) \equiv 0(4)$ M <u>cannot be immersed in</u> $S^{8q-2\alpha(q)-1}$ <u>and cannot be embedded in</u> $S^{8q-2\alpha(q)}$.
2) <u>if</u> $\alpha(q) \equiv 1(4)$ M <u>cannot be immersed in</u> $S^{8q-2\alpha(q)}$.
3) <u>if</u> $\alpha(q) \equiv 2(4)$ M <u>cannot be immersed in</u> $S^{8q-2\alpha(q)+1}$.
4) <u>if</u> $\alpha(q) \equiv 3(4)$ M <u>cannot be immersed in</u> $S^{8q-2\alpha(q)+1}$ <u>and cannot be embedded in</u> $S^{8q-2\alpha(q)+2}$.

This corollary contains as a special case all of the best known non-immersion and non-embedding results for CP^{2q} cf. James [5].

§4 <u>Applications to riemannian geometry</u>

We now turn to the subject of minimal submanifolds. We take the point of view that we are given a fixed riemannian manifold $\bar{M}$ and we are interested in studying questions of existence and structure of minimal submanifolds in $\bar{M}$. Such questions are of particular importance when $\bar{M}$ is one of the classical symmetric spaces such as S^N, $\mathbb{R}P^N$, CP^N, etc., where the connection and curvature are easily computable.

Suppose then that M is a compact, oriented n-dimensional submanifold of $\bar{M}$, and let $\overline{T}$ denote the tangent bundle of $\overline{M}$. Restricting $\overline{T}$ to M we have the natural splitting given by the metric:

$$\overline{T}|_M = T \oplus N$$

where T and N are the tangent and normal bundles of M. The connection on $\overline{M}$ restricts to a connection ∇ on $\overline{T}|_M$ which in general does not preserve the splitting. Nevertheless, we can define a Dirac operator $\overline{D}$ on $Cl(T \oplus N)$, or any bundle of spinors for $T \oplus N$, by the formula

$$\overline{D} = \sum_{j=1}^{n} e_j \cdot \nabla_{e_j}$$

where as before $(e_1, \ldots, e_n)$ is any local orthonormal tangent frame on M. We also have well defined volume elements:

$$\omega_T = e_1 \cdots e_n$$

$$\omega_N = \nu_1 \cdots \nu_2$$

where $(e_1, \ldots, e_n)$ and $(\nu_1, \ldots, \nu_q)$ are oriented orthonormal frames for T and N respectively.

<u>Theorem 4.1</u>. <u>The following are equivalent.</u>

(i) M <u>is a minimal submanifold of</u> $\overline{M}$.

(ii) $\overline{D}$ <u>is self-adjoint</u>.

(iii) $\overline{D}(\omega_T) = 0$

(iv) $\overline{D}(\omega_N) = 0$

It follows that for minimal submanifolds we get a new self-adjoint elliptic operator, and certain natural complexes are defined.

In particular, if M is $4q$-dimensional and spin, we can consider the bundle S of spinors associated to $T \oplus N$. The kernel of $\overline{D}$ on S will be called the space of <u>ambient harmonic spinors</u>.

Theorem 4.2. *Suppose M^{4q} is a compact spin manifold minimally immersed in S^{4q+k} with second fundamental form B satisfying the inequality*

$$\|B\|^2 < \frac{q(4q-1)}{2\sqrt{k+1}+1} . \tag{4.1}$$

Then M^{4q} carries no ambient harmonic spinors.

As in Example 2.8 it is now possible to decompose S into bundles $S^+ \oplus S^-$, the $+1$ and -1 eigenspaces of L_{ω_T}. By Theorem 4.1, restriction gives elliptic operators $D^{\pm} : S^{\pm} \to S^{\mp}$ such that $(D^+)^* = D^-$. Using the Atiyah-Singer theorem one can show that the index of D^+ is a positive multiple of $\hat{A}(M)$. Hence, we have the following.

Corollary 4.3. *If M is as above then $\hat{A}(M) = 0$.*

This result is not new since the inequality (4.1) implies that the scalar curvature of M is positive and the vanishing of $\hat{A}(M)$ then follows from results of Lichnerowicz [6].

It should be pointed out that the inequality (4.1) is generally weaker than the one which appears in the work of Simons [8].

We are hopeful that continued investigation in this direction will lead to further interesting results.

§5 Applications to Kähler geometry

We shall now consider complex riemannian manifolds. Let M be such a manifold and let J denote its almost complex structure. We shall assume that J has the property that in each tangent space $T_x(M)$, $J : T_x(M) \to T_x(M)$ is an orthogonal transformation. We shall also assume that the tensor J is parallel in the canonical riemannian connection; this is equivalent to the statement that J commutes with the operation of parallel translation around any closed curve in M. Manifolds with these properties are called Kähler manifolds. They form an important class. Complex projective space with its natural symmetric space metric is Kähler, and any complex submanifold of a Kähler manifold is Kähler in the induced metric. Hence, the class contains all projective algebraic manifolds.

Suppose then that M is a compact Kähler manifold of complex dimension n, and consider the tangent bundle T of M as a real $2n$-dimensional vector bundle. We then form the associated bundle $Cl(M)$, of real Clifford algebras and take its complexification

$$\mathbb{C}l(M) = Cl(M) \otimes \mathbb{C}$$

At any point $x \in M$, the linear map $J : T_x(M) \to T_x(M)$ extends naturally to all of $Cl(M)_x$, and therefore to $\mathbb{C}l(M)_x$, as a derivation.

For any $x \in M$, we can choose an orthonormal basis of $T_x(M)$ of the form: $e_1, Je_1, \ldots, e_n, Je_n$. In terms of these we define a new basis $\varepsilon_1, \ldots, \varepsilon_n$, $\bar{\varepsilon}_1, \ldots, \bar{\varepsilon}_n$ of $T_x(M) \otimes \mathbb{C}$ (over $\mathbb{C}$) by setting

$$\varepsilon_j = \frac{1}{2}(e_j - \iota Je_j)$$

$$\bar{\varepsilon}_j = \frac{1}{2}(e_j + \iota Je_j)$$

for $j = 1, \ldots, n$. These elements have the property that $J(\varepsilon_j) = \iota\varepsilon_j$ and $J(\bar{\varepsilon}_j) = -\iota\bar{\varepsilon}_j$.

Clearly the elements ε_j and $\bar{\varepsilon}_j$ generate the Clifford algebra $\mathbb{C}l(M)_x$. In fact this algebra has an additive basis of the form $\varepsilon_I\bar{\varepsilon}_{I'}$, where I and I' range over all strictly increasing multi-indices of length $\leq n$. Since J acts as a derivation, one easily sees that

$$J(\varepsilon_I\bar{\varepsilon}_{I'}) = \iota(|I| - |I'|)\varepsilon_I\bar{\varepsilon}_{I'} .$$

Hence, the eigenvalues of J acting on $\mathbb{C}l(M)_x$ are precisely $-\iota n, -\iota(n-1), \ldots, \iota n$. This gives a decomposition of the bundle:

$$\mathbb{C}l(M) = \bigoplus_{p=-n}^{n} \mathbb{C}l^p(M)$$

into eigenbundles for J.

The algebra $Cl(M)_x$ is generated by the ε_j's, $\bar{\varepsilon}_j$'s and 1. The relations are that any pair of elements from $\varepsilon_1, \ldots, \varepsilon_n, \bar{\varepsilon}_1, \ldots, \bar{\varepsilon}_n$ anticommute except pairs of the form $\varepsilon_j, \bar{\varepsilon}_j$. For these we have that $\varepsilon_j\bar{\varepsilon}_j + \bar{\varepsilon}_j\varepsilon_j = -1$.

Now we are able to define three natural operators $\mathcal{L}, \bar{\mathcal{L}}$ and $\mathcal{H}$ on the bundle $Cl(M)$ as follows. For any $\varphi \in Cl(M)_x$,

$$\mathcal{L}(\varphi) = -\sum_{j=1}^{n} \varepsilon_j \varphi \bar{\varepsilon}_j$$

$$\bar{\mathcal{L}}(\varphi) = -\sum_{j=1}^{n} \bar{\varepsilon}_j \varphi \varepsilon_j$$

$$\mathcal{H} = [\mathcal{L}, \bar{\mathcal{L}}] .$$

These operators are independent of the choice of unitary basis $\varepsilon_1, \ldots, \varepsilon_n$. It is not difficult to see that $\mathcal{H}(\varphi) = w\varphi + \varphi w + n\varphi$ where $w = \Sigma \bar{\varepsilon}_j \varepsilon_j$. The basic fact concerning these operators is that they satisfy the following relations.

<u>Proposition 5.1.</u>

$$[\mathcal{H}, \mathcal{L}] = 2\mathcal{L}$$

$$[\mathcal{H}, \bar{\mathcal{L}}] = -2\bar{\mathcal{L}}$$

$$[\mathcal{L}, \bar{\mathcal{L}}] = \mathcal{H}$$

We thereby have a natural $\mathfrak{sl}_2(\mathbb{C})$-structure defined on $Cl(M)$. Each of these operators commutes with J and hence preserves the subbundles $Cl^p(M)$. The semi-simple element $\mathcal{H}$ has integer eigenvalues. Taking the corresponding eigenbundles we get a further decomposition

$$Cl(M) = \bigoplus_{p,q} Cl^{p,q}(M) .$$

Under the canonical isomorphism $Cl(M) \cong \Lambda^*(M) \otimes \mathbb{C}$. This decomposition is definitely different from the Dolbeault decomposition.

For a given $p, -n \leq p \leq n$, only certain of the bundles $Cl^{p,q}(M)$ can be non-zero. In fact, $Cl^{p,q}(M)$ is non-zero only if $|p+q| \leq n$ and $p+q+n$ is even. Hence, the bundles appear only for values of (p,q) marked in the "diamond" pictured on the last page.

We now define differential operators on sections of the Clifford bundle as follows:

$$\mathcal{D} = \sum_{j=1}^{n} \varepsilon_j \cdot \nabla_{\bar{\varepsilon}_j}$$

$$\bar{\mathcal{D}} = \sum_{j=1}^{n} \bar{\varepsilon}_j \cdot \nabla_{\varepsilon_j}$$

where the ε_j's are as before and where ∇ denotes the riemannian connection. These operators are again independent of the choice of the ε_j's. They can be related to the real Dirac operator D as follows.

$$D = \sum e_j \cdot \nabla_{e_j} + Je_j \cdot \nabla_{Je_j} = \sum \eta_j \cdot \nabla_{\eta_j}$$

where $\eta_1, \dots, \eta_{2n}$ is any orthonormal basis of $T_x(M)$. We define a conjugate Dirac operator D^c by setting

$$D^c = \sum e_j \cdot \nabla_{Je_j} - Je_j \cdot \nabla_{e_j} = \sum J\eta_j \cdot \nabla_{\eta_j} .$$

Then

$$\mathcal{D} = \frac{1}{4}(D - iD^c)$$

$$\bar{\mathcal{D}} = \frac{1}{4}(D + iD^c)$$

These differential operators satisfy certain simple relations with $\mathcal{L}, \bar{\mathcal{L}}$ and $\mathcal{H}$ which allow one to prove that

5.1 $$\mathcal{D} : \Gamma Cl^{p,q}(M) \to \Gamma Cl^{p+1,q+1}(M)$$

$$\overline{\mathcal{D}} : \Gamma Cl^{p+1,q+1}(M) \to \Gamma Cl^{p,q}(M) \tag{5.2}$$

for all p, q. Furthermore, $\mathcal{D}$ and $\overline{\mathcal{D}}$ are adjoints of one another and have the property that

$$\mathcal{D}^2 = \overline{\mathcal{D}}^2 = 0$$

Consequently, we can define a sequence of complexes:

$$\cdots \xrightarrow{\mathcal{D}} \Gamma Cl^{p-1,q-1}(M) \xrightarrow{\mathcal{D}} \Gamma Cl^{p,q}(M) \xrightarrow{\mathcal{D}} \Gamma Cl^{p+1,q+1}(M) \xrightarrow{\mathcal{D}} \cdots \tag{5.3}$$

with associated Clifford cohomology groups $\mathcal{H}^{p,q}$.

One can show that the complex (5.3) is elliptic; that is, the symbol sequence is exact. It follows therefore from general theory that there is a decomposition

$$\Gamma(Cl^{p,q}(M)) = \mathbb{H}^{p,q} \oplus \operatorname{Im}(\mathcal{D}) \oplus \operatorname{Im}(\overline{\mathcal{D}})$$

where

$$\begin{aligned} \mathbb{H}^{p,q} &= \ker(\mathcal{D}\overline{\mathcal{D}} + \overline{\mathcal{D}}\mathcal{D}) \quad \text{on} \quad Cl^{p,q}(M) \\ &= \ker \mathcal{D} \cap \ker \overline{\mathcal{D}} \quad \text{on} \quad Cl^{p,q}(M) \end{aligned}$$

and

$$\mathcal{H}^{p,q} \cong \mathbb{H}^{p,q}$$

It also follows that the groups $\mathcal{H}^{p,q}$ are finite dimensional. They do not correspond to the Dolbeault groups under the canonical isomorphism.

The complexes $(Cl^{*,*}(M), \mathcal{D})$ can be represented schematically by placing a dot at each point in the (p,q) plane such that $Cl^{p,q}(M)$ is non-zero. The operators $\mathcal{D}$ and $\overline{\mathcal{D}}$ then shift along the diagonal lines $p - q = \text{constant}$. To complete the symmetry of the picture we note that a pair of analogous operators $\mathcal{D}^{\wedge}$ and $\overline{\mathcal{D}}^{\wedge}$ can be defined using right Clifford multiplication as follows.

$$\mathcal{D}^{\wedge}(\varphi) = \sum \left(\nabla_{\overline{\epsilon}_j}\varphi\right)\epsilon_j \quad \text{and} \quad \overline{\mathcal{D}}^{\wedge} = \sum \left(\nabla_{\epsilon_j}\varphi\right)\overline{\epsilon}_j .$$

This pair of operators has properties similar to those of the pair $\mathcal{D}$ and $\overline{\mathcal{D}}$; however,

$$\mathcal{D}^{\wedge} : \Gamma Cl^{p,q}(M) \to \Gamma Cl^{p+1,q-1}(M) \quad \text{and} \quad \overline{\mathcal{D}}^{\wedge} : \Gamma Cl^{p+1,q-1}(M) \to \Gamma Cl^{p,q}(M) .$$

The corresponding harmonic spaces $\hat{\mathbb{H}}^{p,q}$ for this pair can be shown to coincide with $\mathbb{H}^{p,q}$. Hence, the corresponding cohomology groups are the same.

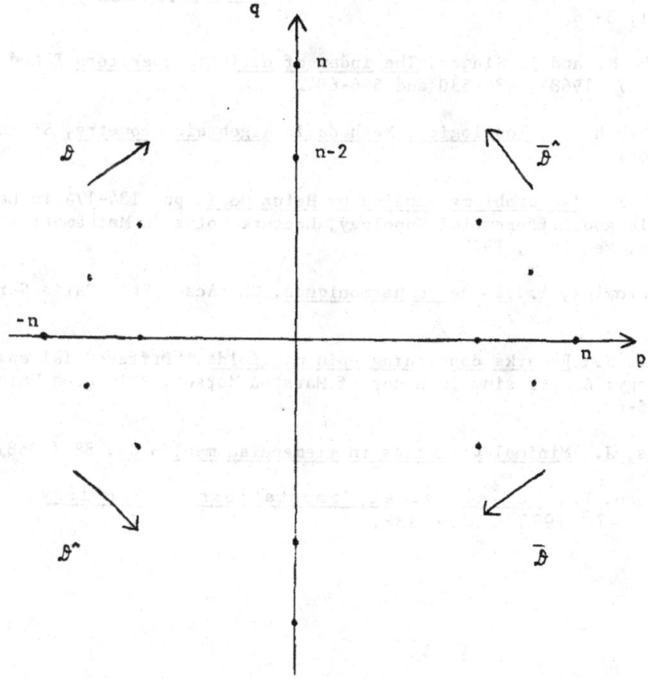

Finally, one can prove the following result.

<u>Theorem 5.2</u>. <u>The operators</u> $\mathcal{L}, \bar{\mathcal{L}}$ <u>and</u> $\mathcal{H}$ <u>descend to Clifford cohomology and satisfy the same commutation relations.</u>

BIBLIOGRAPHY

[1] Atiyah, M., Vector fields on Manifolds, Arbeitsgemeinschaft für Forschung des Landes Nordrhein-Westfalen, Heft 200.

[2] Atiyah, M., Bott, R. and A.A. Shapiro, Clifford modules, Topology 3 (Suppl. 1) (1964), 3-38.

[3] Atiyah, M. and I. Singer, The index of elliptic operators I and III, Ann. of Math. 87 (1968), 484-530 and 546-604.

[4] Hirzebruch, F., Topological Methods in Algebraic Geometry, Springer-Verlag, New York.

[5] James, I., Two problems studied by Heinz Hopf, pp. 134-174 in Lectures on Algebraic and Differential Topology, Lecture Notes in Mathematics 279, Springer-Verlag, New York, 1972.

[6] Lichnerowicz, A., Spineurs harmoniques, CR. Acad. Sci. Paris Ser. A-B 257(1963) 7-9.

[7] Milnor, J., Remarks concerning spin manifolds, "Differential and Combinatorial Topology: A Symposium in honor of Marston Morse", Princeton Univ. Press (1965), pp. 55-62.

[8] Simons, J., Minimal varieties in riemannian manifolds, 88 (1968), 62-105.

[9] Sullivan, D., Infinitesimal Computations in Topology, IHES Pub. Math. 47 (1977), 269-332.

SEIFERT MANIFOLDS, PLUMBING, μ-INVARIANT AND ORIENTATION REVERSING MAPS

by Walter D. Neumann* and Frank Raymond*
The University of Maryland and the University of Michigan

Dedicated to R.L. Wilder

Homology spheres and the μ-invariant appear to play a crucial role in many problems in low dimensional topology. For example, the existence of a 3-dimensional homology sphere Σ of μ-invariant 8 for which $\Sigma \# \Sigma$ bounds an acyclic 4-manifold would imply triangulability of all manifolds of dimension ≥ 6. As Casson has remarked, if Σ has μ-invariant 8 and admits an orientation reversing homeomorphism, then Σ would satisfy this criterion.

As another example, the intersection form of a simply connected almost parallelizable 4-manifold is even. No such closed 4-manifold with non-trivial definite form is known. A reasonable and popular procedure to collect empirical data towards the existence or non-existence of such manifolds is to study 4-manifolds with $(Z/2)$-homology sphere boundaries, since pasting along such boundaries preserves evenness of forms.

In this paper we have compiled some results and computations in these areas for the special case of Seifert 3-manifolds and other plumbed manifolds. For example, we classify in section 8 those Seifert manifolds which admit orientation reversing homeomorphisms. No rational homology spheres other than lens spaces are among them.

Section 1 reviews the fundamentals of Seifert manifolds in a more convenient version than the usual one.

We show in sections 2, 3, and 4 that the class of Seifert manifolds which are homology spheres coincides precisely with a natural subclass of the class of Brieskorn complete intersections (studied in section 2) and also with a natural subclass of the class of homogeneous spaces discussed in section 3. Thus Seifert homology spheres arise as links of isolated complex surface singularities with $\mathbb{C}^*$-action. We show, in fact, in section 5 that almost every Seifert manifold arises

*Research supported in part by the National Science Foundation.

this way, namely precisely those which do not fiber equivariantly over S^1. We do this by describing a "canonical plumbing diagram" for such Seifert manifolds. We give a table of Seifert homology spheres with small invariants which can be made to bound even definite 4-manifolds by these methods.

In sections 6 and 7 we give various useful algorithms for computing μ-invariants for Seifert $(Z/2)$-homology spheres and other plumbed manifolds. A table is included.

1. Fundamentals

In this paper, with the exception of §8, "Seifert manifold" will always mean an oriented closed connected 3-manifold admitting a fixed point free action of S^1. Such a manifold is equivariantly classified by its "Seifert invariants" [S],[OR]. We shall use non-normalized Seifert invariants in this paper (see [N1]) since they are more convenient for calculations. They are described as follows.

Let $M^3 \to M^3/S^1 = F$ be the Seifert fibration. Let $O_1, \ldots, O_s$ be a non-empty collection of disjoint orbits in M, including all singular orbits. Let $T_1, \ldots, T_s$ be disjoint invariant tubular neighborhoods of $O_1, \ldots, O_s$ and $M_0 = M - \text{int}(T_1 \cup \ldots \cup T_s)$. Since $M_0 \to M_0/S^1 = F_0$ is an S^1-bundle over a connected surface with boundary, it admits a section $R \subset M_0$. Let $R_i = R \cap \partial T_i$. After choosing orientation conventions, R_i is a curve in ∂T_i which is homologous in T_i to some multiple $\beta_i O_i$ of the central curve. Let α_i be the order of the isotropy subgroup $Z/\alpha_i \subset S^1$ at the orbit O_i. Let g be the genus of the surface F. Then the unnormalized Seifert invariant is the collection of numbers

$$(g\,;\,(\alpha_1,\beta_1),\ldots,(\alpha_s,\beta_s))\,.$$

They satisfy $g \geq 0$, $\alpha_i \geq 1$, $\gcd(\alpha_i,\beta_i) = 1$.

The Seifert invariant is not unique: we can add or remove principal orbits from our collection of orbits O_i and we can choose different sections $R \subset M_0$. The following theorem is easily proved ([N1]).

<u>Theorem 1.1</u>. <u>Let M and M' be two Seifert manifolds with associated Seifert in-</u>

variants $(g\,;(\alpha_1,\beta_1),\ldots,(\alpha_s,\beta_s))$ and $(g';(\alpha'_1,\beta'_1),\ldots,(\alpha'_t,\beta'_t))$ respectively. Then M and M' are orientation preservingly homeomorphic by a fiber preserving homeomorphism if and only if, after reindexing the Seifert pairs if necessary, there exists a k such that

(i) $\alpha_i = \alpha'_i$ for $i = 1, \ldots, k$ and $\alpha_i = \alpha'_j = 1$ for $i, j > k$.

(ii) $\beta_i \equiv \beta'_i \pmod{\alpha_i}$ for $i = 1, \ldots, k$.

(iii) $\Sigma_{i=1}^{s} (\beta_i/\alpha_i) = \Sigma_{i=1}^{t} (\beta'_i/\alpha'_i)$.

Remark. It is easy to check that (i), (ii), (iii) above are equivalent to: $(g',(\alpha'_j,\beta'_j), j=1,\ldots,t)$ can be obtained from $(g,(\alpha_i,\beta_i), i = 1,\ldots,s)$ by a sequence of the following moves:

a) permute the indices;

b) add or delete a Seifert pair $(1,0)$;

c) replace (α_1,β_1), (α_2,β_2) by $(\alpha_1,\beta_1+m\alpha_1)$, $(\alpha_2,\beta_2-m\alpha_2)$ for some $m \in Z$.

Definition. Denote the number $-\Sigma_{i=1}^{s}(\beta_i/\alpha_i)$, which is an invariant of the Seifert manifold by (iii), by $e(M)$, called the Euler number of M. We assume we have chosen our orientation conventions earlier, at the point where we were not specific about them, so that $e(M)$ is the usual Euler number if M is an S^1-bundle.

Note that reversing the orientation of M, either by reversing the orientation of the fibers or of the base (it does not matter which, since M admits an orientation preserving self-homeomorphism mapping fibers to fibers and reversing orientation both on fibers and base), replaces the Seifert invariant $(g,(\alpha_i,\beta_i))$ by $(g,(\alpha_i,-\beta_i))$. Hence $e(-M) = -e(M)$.

The Euler number has a delightful naturality property, which is invaluable for computations, as we shall see later.

Theorem 1.2. Let M and M' be Seifert manifolds with base spaces F and F', and let $f: M \to M'$ be an orientation preserving fiber preserving homeomorphism. Let the degree of the induced map on a typical fiber be n and the degree of the induced map $\overline{f}: F \to F'$ be m. Then $e(M) = (m/n)e(M')$.

Here is the idea behind the theorem. If $M \to F$ and $M' \to F'$ were genuine

S^1-bundles, the theorem would be an easy cohomology calculation, since $e(M) = \langle c(M),[F]\rangle$ where $c(M) \in H^2(F;Z) = [F,K(Z,2)] = [F,BS^1]$ is given by the classifying map $F \to BS^1$ of the bundle. Although $M \to F$ is not a genuine bundle, we can make it into a genuine bundle by replacing the fibers by the "rationalized circle" $S^1_{(0)}$. This replaces BS^1 by $BS^1_{(0)} = K(\mathbb{Q},2)$, so the argument sketched above then goes through. To make this argument precise, it is easiest to observe[1)] that we do not have to "localize all the way to $\mathbb{Q}$". Let a be a positive integer divisible by all the α_i's occurring in M or M'. Factor by the (Z/a)-action inside the S^1-action to get a diagram of maps

$$\begin{array}{ccc} M & \xrightarrow{\quad f \quad} & M' \\ \Big\downarrow{\scriptstyle \pi} & & \Big\downarrow{\scriptstyle \pi'} \\ M/(Z/a) & \xrightarrow{\quad f_a \quad} & M'/(Z/a) \end{array}$$

We need that f can be made (Z/a)-equivariant, but this is easily done. Now $M/(Z/a)$ and $M'/(Z/a)$ are both genuine S^1-bundles, so the theorem is true for f_a. If we show it is true for π and π', then it follows for f. But for π and π' it follows from the following lemma (due to Seifert [S]).

Lemma 1.3. _If M has Seifert invariants $(g\,;(\alpha_i,\beta_i))$ then $M/(Z/a)$ has Seifert invariants $(g\,;(\bar{\alpha}_i,\bar{\beta}_i))$ where $\bar{\beta}_i/\bar{\alpha}_i$ is $a\beta_i/\alpha_i$ expressed in lowest terms._

Proof. The section $R \subset M_0 = M - \mathrm{int}(T_1 \cup \ldots \cup T_s)$ used to compute the (α_i,β_i) projects down to a section $\bar{R} \subset M_0/(Z/a)$, which, when used to compute the Seifert invariant of $M/(Z/a)$, yields the lemma.

Theorem 1.1 says that a Seifert manifold M is determined by knowledge of its Euler number $e(M)$ and by knowledge of the pairs $(\alpha_i,\beta_i \bmod \alpha_i)$. But $(\alpha_i,\beta_i(\mathrm{mod}\ \alpha_i))$ is equivalent to knowing the "slice type" of the corresponding orbit, that is, the equivalence class of the representation of the isotropy subgroup in the normal plane to the orbit. Thus if $M \to M'$ is a branched covering of Seifert

1) This observation is due to Howard Rees.

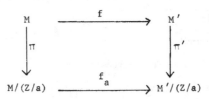

manifolds, coming maybe from factoring by a group action, then $e(M)$ and $e(M')$ determine each other by Theorem 1.2 while the slice types of M can generally be computed from those for M' and vice versa by elementary local computations.

In the following, we shall often write $M = M(g\,;(\alpha_i,\beta_i), i=1,\dots,m)$ as an abbreviation for "M has Seifert invariant $(g\,;(\alpha_i,\beta_i), i=1,\dots,m)$".

2. Examples; Brieskorn Complete Intersections

Let $a_1, \dots, a_n$ be integers, $a_i \geq 1$. Then if $A = (\alpha_{ij})$ is a sufficiently general $(n-2)\times n$ - matrix of complex numbers, the variety

$$V_A(a_1,\dots,a_n) = \{z \in \mathbb{C}^n \mid \alpha_{i1}z_1^{a_1} + \dots + \alpha_{in}z_n^{a_n} = 0, i=1,\dots,n-2\}$$

is a complex surface which is non-singular except perhaps at the origin and

$$\Sigma^3(a_1,\dots,a_n) = V_A(a_1,\dots,a_n) \cap S^{2n-1}$$

is a smooth 3-manifold which does not depend on A up to diffeomorphism. A is in fact sufficiently general if (and if all $a_i \geq 2$ also only if) every $(n-2)\times(n-2)$ subdeterminant of A is nonzero, by Hamm [Ha]. We assume A satisfies this from now on.

$V_A(a_1,\dots,a_n)$ has a $\mathbb{C}^*$-action by $t(z_1,\dots,z_n) = (t^{q_1}z_1,\dots,t^{q_n}z_n)$ for $t \in \mathbb{C}^*$, where $q_j = (\mathrm{lcm}\, a_i)/a_j$. $S^1 \subset \mathbb{C}^*$ acts fixed point freely on $\Sigma(a_1,\dots,a_n)$, so $\Sigma(a_1,\dots,a_n)$ is a Seifert manifold.

__Theorem 2.1.__ $\Sigma = \Sigma(a_1,\dots,a_n)$ __has Seifert invariants__ $(g\,; s_1(t_1,\beta_1),\dots,s_n(t_n,\beta_n))$, __where__ $s_j(t_j,\beta_j)$ __means__ (t_j,β_j) __repeated__ s_j __times and__

$$t_j = \mathrm{lcm}_i(a_i)/\mathrm{lcm}_{i\neq j}(a_i)$$

$$s_j = \prod_{i\neq j}(a_i)/\mathrm{lcm}_{i\neq j}(a_i)$$

$$g = \tfrac{1}{2}\Big(2 + (n-2)\prod_i(a_i)/\mathrm{lcm}(a_i) - \sum_j s_j\Big)$$

__and the__ β_i __and the Euler number__ $e(\Sigma)$ __are determined (up to equivalence of Seifert__ invariants) by

$$-e(\Sigma) = \sum s_j \frac{\beta_j}{t_j} = \prod (a_i)/(\operatorname{lcm} a_i)^2 .$$

Note that the latter equation can be rewritten (by dividing through by its right hand side)

$$\sum \beta_j q_j = 1 , \quad \text{where} \quad q_j = \operatorname{lcm}(a_i)/a_j .$$

But clearly t_i divides q_j if $i \neq j$ and is prime to q_j if $i = j$. Thus modulo t_j the equation becomes $\beta_j q_j \equiv 1 \pmod{t_j}$ and hence determines $\beta_j \pmod{t_j}$, as claimed in the theorem.

Proof of theorem. First note that the only points of $\Sigma = \Sigma^3(a_1,\dots,a_n)$ with non-trivial isotropy are points with some coordinate zero. The condition on the coefficient matrix A implies that $V_A \cap \{z_i = z_j = 0\} = \{0\}$ for $i \neq j$, so we need only consider $z \in \Sigma^3$ with one coordinate zero, say $z_j = 0$. At such a point the isotropy subgroup has order $\gcd_{i\neq j}(q_i) = t_j$. An easy counting argument (see [N1]) can be used to see that $\Sigma^3 \cap \{z_j = 0\}$ consists of exactly s_j orbits, but this follows also from the later discussion.

Observe that we can write

$$\Sigma^3(a_1,\dots,a_n) = (V_A(a_1,\dots,a_n) - \{0\})/\mathbb{R}_+ ,$$

where $\mathbb{R}_+ \subset \mathbb{C}^*$ is in the $\mathbb{C}^*$-action. Denote $(V_A - \{0\})/\mathbb{C}^* = \Sigma/S^1$ by $P(a_1,\dots,a_n)$. Consider the diagram

$$\begin{array}{ccc}
V_A(a_1,\dots,a_n) - \{0\} & \xrightarrow{\Phi} & V_A(1,\dots,1) - \{0\} \\
\downarrow /\mathbb{R}_+ & & \downarrow /\mathbb{R}_+ \\
\Sigma(a_1,\dots,a_n) & \xrightarrow{\varphi} & \Sigma(1,\dots,1) \\
\downarrow /S^1 & & \downarrow /S^1 \\
P(a_1,\dots,a_n) & \xrightarrow{\bar{\varphi}} & P(1,\dots,1)
\end{array}$$

with horizontal arrows induced by $\Phi(z_1,\dots,z_n) = (z_1^{a_1},\dots,z_n^{a_n})$. We intend to apply

Theorem 1.2 to the map φ to compute $e(\Sigma)$.

Note that Φ and φ are equivariant if we let C^* and S^1 act non-effectively on $V_A(1,\ldots,1)$ and $\Sigma(1,\ldots,1)$ by $t(z_1,\ldots,z_n) = (t^a z_1,\ldots,t^a z_n)$, $a = \mathrm{lcm}(a_i)$. Thus φ has degree a on a typical fiber.

To determine the degree of $\bar{\varphi}$, note that the group $H = (Z/a_1)\times\ldots\times(Z/a_n)$ acts on each space on the left of the diagram by letting Z/a_j act by multiplication by $e^{2\pi i/a_j}$ in the j-th coordinate. The map Φ can be identified with the orbit map $V_A(a_1,\ldots,a_n)-\{0\} \to (V_A(a_1,\ldots,a_n)-\{0\})/H$, and similarly for φ and $\bar{\varphi}$. Considering S^1 and H both as subgroups of $\mathrm{Diff}(\Sigma(a_1,\ldots,a_n))$ by these actions, denote $H_0 = S^1 \cap H$. Now on the one hand, H_0 is isomorphic to the non-effectivity kernel of S^1 acting on $\Sigma(a_1,\ldots,a_n)/H = \Sigma(1,\ldots,1)$, so $H_0 \cong Z/a$, while on the other hand H_0 is the non-effectivity kernel of H acting on $\Sigma(a_1,\ldots,a_n)/S^1 = P(a_1,\ldots,a_n)$, so the orbit map $\bar{\varphi}$ of this action has degree $|H/H_0| = \Pi(a_i)/a$.

Now $V_A(1,\ldots,1) \subset C^n$ is a linear subspace and hence $\Sigma(1,\ldots,1) \to P(1,\ldots,1)$ is the usual Hopf map $S^3 \to CP^1$. Thus $e(\Sigma(1,\ldots,1)) = -1$, so by Theorem 1.2, $e(\Sigma(a_1,\ldots,a_n)) = -\Pi(a_i)/a^2$.

Finally to compute g , note that the subspace $z_j = 0$ of $P(1,\ldots,1)$ is a single point and that these points are precisely the points where branching of Φ occurs. The argument we used to show $\bar{\varphi}$ itself has degree $\Pi a_i/\mathrm{lcm}\, a_i$ applies with one coordinate less to show that $\bar{\varphi}$ restricted the subspaces $z_j = 0$ of $P(a_1,\ldots,a_n)$ and $P(1,\ldots,1)$ has degree $\Pi_{i\neq j} a_i/\mathrm{lcm}_{i\neq j} a_i = s_j$, so $P(a_1,\ldots,a_n)$ contains exactly s_j points with $z_j = 0$ (proving, by the way, that $\Sigma(a_1,\ldots,a_n) \cap \{z_j = 0\}$ consists of s_j orbits, as promised earlier). The standard formula for euler characteristic of a branched cover thus gives

$$\chi(P(a_1,\ldots,a_n)) = \left(\prod(a_i)/a)\chi(P(1,\ldots,1)\right) + \sum_j (s_j - \prod(a_i)/a)$$

$$= (2-n)\prod_i (a_i)/a + \sum_j s_j ,$$

yielding the value for g claimed in the theorem.

<u>Remark</u>. One can also give a very elementary computation of the Seifert pairs

(t_j,β_j) and the Euler number $e(\Sigma)$ by observing that if the β_j are chosen to satisfy $\Sigma\beta_j q_j = 1$, then

$$R = \{z \in \Sigma \mid z_j = r_j e^{i\theta_j}, r_j > 0, \sum \beta_j\theta_j \equiv 0 \pmod{2\pi}\}$$

is a section to the S^1-action in the complement of the exceptional orbits which yields, via the definition of the Seifert invariant, the values (t_j,β_j) for the Seifert pairs. However, one still needs a computation like the above proof to determine g.

A completely analogous proof to the above shows more generally

Theorem 2.2. Let $a_1, \ldots, a_n$, $d_1, \ldots, d_{n-2}$ be positive integers and $\gcd(d_i) = 1$. Let

$$V = \{z \in \mathbb{C}^n \mid \alpha_{i1}z_1^{d_i a_1} + \cdots + \alpha_{in}z_n^{d_i a_n} = 0, i = 1,\ldots,n-2\}$$

with sufficiently general coefficients α_{ij}, and let $\Sigma = V \cap S^{2n-1}$. Then Σ is a Seifert manifold with invariant $(g\,;\, ds_j(t_j,\beta_j), j = 1,\ldots,n)$ where t_j and s_j are as in Theorem 2.1, $d = \Pi d_i$,

$$g = 1 - \frac{d}{2}\Big[\sum s_j - \Big(\sum d_i\Big)\prod a_j/\operatorname{lcm} a_j\Big],$$

and $e(\Sigma)$ and the β_j are given by

$$-e(\Sigma) = \sum ds_j \frac{\beta_j}{t_j} = d\prod a_i/(\operatorname{lcm} a_i)^2 .$$

The only alteration necessary in the previous proof is that now $P(1,\ldots,1)$ is replaced by a complete intersection of $n-2$ hypersurfaces of degrees d_i in CP^{n-1}, so it has Euler characteristic $d(n - \Sigma d_i)$ (by the adjunction formula for instance), and by similar reasoning $z_j = 0$ now determines exactly d points in $P(1,\ldots,1)$, instead of just 1.

Bibliographic notes: A general program for computing the Seifert invariant of the link of an isolated surface singularity with $\mathbb{C}^*$-action was given by Orlik and

Wagreich [OW],[O]. The method used here is, however, based on the case $n = 3$ of Theorem 2.1 done by Neumann [N1]. Brieskorn complete intersections of the type in Theorem 2.2 were first introduced by Randell [Ran].

3. Homogeneous Spaces

Using similar methods to the preceding section, one can describe the Seifert invariants of those Seifert manifolds which are homogeneous spaces $\Pi\backslash G$, where Π is a discrete subgroup of a Lie group G with compact quotient. This was done by Raymond and Vasquez [RV]. We describe the result for $G = \widetilde{PSL}(2;\mathbb{R})$, the universal cover of $PSL(2,\mathbb{R})$.

Theorem 3.1. *$M((g),(\alpha_i,\beta_i),i=1,\ldots,s)$ has the form $\Pi\backslash G$ if and only if there exists a divisor q of $\alpha_1 \ldots \alpha_s(g+s-2-\Sigma 1/\alpha_i)$ prime to each α_i such that $\beta_i q \equiv -1 \bmod \alpha_i$ for each i and*

$$e(M) = -\frac{1}{q}(g+s-2-\sum 1/\alpha_i)\,.$$

In this case Π is a subgroup of index q in the group $\Gamma = \pi^{-1}Q$, where $\pi : G \to PSL(2,\mathbb{R})$ is the covering, and $Q = \pi(\Gamma)$ is a Fuchsian group with signature $(g;\alpha_1,\ldots,\alpha_s)$.

Remark 3.2. There are precisely q^{2g} subgroups $\Pi\subset\pi^{-1}(Q)$ of index q with $\pi(\Pi) = Q$. They are all related by automorphisms of $\pi^{-1}(Q)$. This gives a classification of discrete subgroups $\Pi \subset G$ with $\Pi\backslash G$ compact.

Example 3.3. It is easy to apply this to Theorem 2.1 to see that $\Sigma(a_1,\ldots,a_n)$ has the form $\Pi\backslash G$. In this case Π is the commutator subgroup of $\Gamma = \pi^{-1}(Q)$, where $Q \subset PSL(2,\mathbb{R})$ is a Fuchsian group of signature $(0;a_1,\ldots,a_n)$. This can be shown by explicit computation of Seifert invariants, which was how we originally did it. Using automorphic forms, one can prove a stronger version of the same result ([N2]). This has been done independently by I. Dolgačev. The case $n = 3$ was done by J. Milnor [M], and F. Klein [K] for $G = SU(2)$. More generally, the manifold of Theorem 2.2 has the form $\pm\Pi\backslash G$ if and only if

$$q = \pm(1/d)\,\mathrm{lcm}\, a_i\Big[\sum d_j - d\sum 1/a_i\Big], \quad (d = \prod d_j),$$

is a positive integer and is prime to $\alpha_j = \mathrm{lcm}\, a_i/\mathrm{lcm}_{i\neq j} a_i$ for each j. This holds for, example, if d divides Σd_j.

4. Homology Spheres

Theorem 4.1. If the Seifert manifold $M = M(g\,;(\alpha_i,\beta_i), i=1,\dots,m)$ *is a* Z-*homology sphere, then* $g = 0$ *and the* α_i *are pairwise coprime. Furthermore, to given pairwise coprime* α_i *there is exactly one* Z-*homology sphere as above, up to orientation. It is diffeomorphic to the Brieskorn complete intersection manifold* $\Sigma(\alpha_1,\dots,\alpha_m)$, *and hence also to a homogeneous space* $\Pi\backslash G$ *as in Example* 3.3.

Remark. It follows that the subgroups $\Pi \subset G$ of Example 3.3 corresponding to pairwise coprime exponents $a_1, \dots, a_n$ are the only discrete $\Pi \subset G$ with $\Pi\backslash G$ compact for which Π is perfect (i.e., $[\Pi,\Pi] = \Pi$).

Proof of theorem. The first two sentences of the theorem are due to Seifert [S]. Namely, if M is as above, then by abelianizing the standard presentation of $\pi_1(M)$, Seifert showed $H_1(M) \cong Z^{2g} \oplus \mathrm{Cok}(A)$, where $A : Z^m \to Z^m$ is a map with matrix

$$A = \begin{pmatrix} 1 & 1 & \cdots & 1 & 0 \\ \alpha_1 & 0 & \cdots & 0 & \beta_1 \\ & \cdot & & & \\ & & \cdot & & \\ & & & \cdot & \\ & & & & \cdot \\ 0 & 0 & \cdots & \alpha_m & \beta_m \end{pmatrix}.$$

But $\det A = \pm\sum_{i=1}^{m} \beta_i\alpha_1 \cdots \hat{\alpha}_i \cdots \alpha_m = \pm\alpha_1 \cdots \alpha_m \Sigma(\beta_i/\alpha_i)$. The condition that M is a homology sphere is thus: $g = 0$ and $\alpha_1 \cdots \alpha_m \Sigma(\beta_i/\alpha_i) = \pm 1$. By reversing orientation if necessary, we can assume

$$\alpha_1 \cdots \alpha_m \sum (\beta_i/\alpha_i) = +1.$$

This equation implies that the α_i are pairwise coprime. Further, by considering it modulo α_i, we see it determines β_j modulo α_j for each j. It also determines $e(M)$ as

$$e(M) = -\sum \beta_i/\alpha_i = -1/\alpha_1 \cdots \alpha_m .$$

It thus determines M completely for given $\alpha_1, \dots, \alpha_m$.

Comparing with Theorem 2.1 proves the second statement. Alternatively, a simple proof of 2.1 for this case is given by observing that by Hamm [Ha], $\Sigma(\alpha_1,\dots,\alpha_m)$ is a homology sphere for α_i pairwise coprime and its S^1 action clearly has isotropy $Z/\alpha_1, \dots, Z/\alpha_m$.

One can get a simple proof of Example 3.3 for this case also by applying part (iii) of the following lemma.

Lemma 4.2. *If $\alpha_1, \dots, \alpha_m$ are pairwise coprime and $M = M(0,(\alpha_i,\beta_i), i=1,\dots,m)$, then $e(M) = c/\alpha_1 \cdot \ldots \cdot \alpha_m$ for some c prime to $\alpha_1, \dots, \alpha_m$, and*

(i) *M is uniquely determined up to orientation by $|c|$; denote it M_c;*

(ii) *$H_1(M_c) = Z/|c|$, generated by the class of a principal orbit;*

(iii) *M_c covers M_d if and only if d divides c; in particular $M_1 = \Sigma(\alpha_1,\dots,\alpha_n)$ is the maximal abelian cover of M_c for any $c > 0$.*

Proof. Up to and including part (i), this is the same proof as the previous theorem. Parts (ii) and (iii) then follow by observing that $Z/c \subset S^1$ acts freely on M_1, so M_c must be $M_1/(Z/c)$ by Lemma 1.3.

Similar statements hold for $g \neq 0$. We leave their formulation and proof to the reader.

5. Seifert Manifolds via Plumbing

Von Randow's algorithm ([vR], see Orlik [O] or Hirzebruch, Neumann, Koh [HNK] for a description in terms of our present orientation conventions) expressing a Seifert manifold via plumbing extends with no change to unnormalized Seifert invariants, yielding the result:

Theorem 5.1. *Let $M^3 = \partial P(\Gamma)$ be the result of plumbing according to the following*

<u>weighted tree</u>

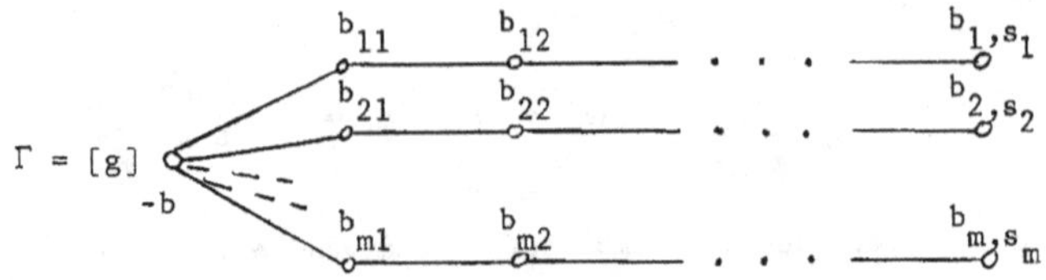

<u>Then</u> $M^3 \cong M((g),(1,b),(\alpha_i,\beta_i),i=1,\dots,m)$, <u>where</u> α_i/β_i <u>is the continued fraction</u>

$$\alpha_i/\beta_i = b_{i1} - 1/(b_{i2} - 1/(b_{i3} - \cdots - 1/b_{is_i})\dots)$$
$$= [b_{i1},\dots,b_{is_i}] \qquad (\underline{\text{notation}}).$$

The [g] above means that the corresponding bundle being plumbed is the bundle of Euler number $-b$ over a surface of genus g ; all the other bundles are bundles of Euler number b_{ij} over the sphere S^2. We omit the [g] if $g = 0$. We are using the notation $P(\Gamma)$ for the four-manifold obtained by plumbing disc bundles according to Γ and $\partial P(\Gamma)$ for its boundary obtained by plumbing circle bundles.

By von Randow [vR], in any plumbing graph Γ we can "blow down" vertices corresponding to a bundle of Euler number $\varepsilon = \pm 1$ over S^2 having at most two adjacent vertices by removing that vertex and replacing the weights b_i of the neighboring vertices of Γ by $b_i - \varepsilon$. This does not change $\partial P(\Gamma)$. For example, if we start with

E_8 =
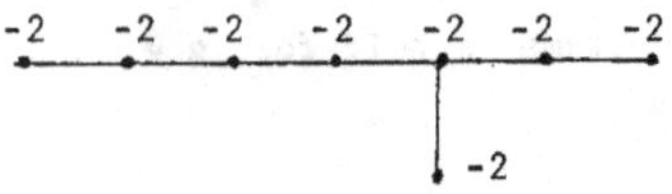

then we can "blow up" (reverse operation of blowing down) to get

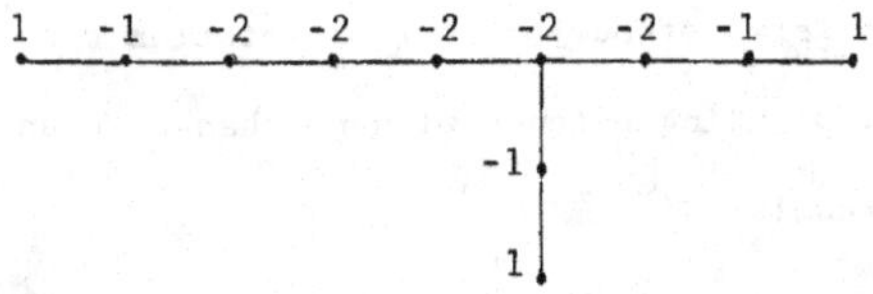

and then by iteratively blowing down -1's we can finally get to

$\Gamma =$ 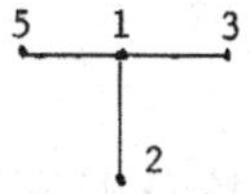

Thus $\partial P(E_8) = \partial P(\Gamma) = \Sigma(2,3,5)$, where the last equality uses Theorem 2.1. It is not hard to show that any two plumbing graphs as in 5.1 for the same Seifert manifold are related by a sequence of blowings up and blowings down.

If Γ is an arbitrary plumbing graph with vertices $v_1, \ldots, v_r$ with Euler number weights $b_1, \ldots, b_r$ and arbitrary genus weights, then the four manifold $P(\Gamma)$ obtained by plumbing disc bundles according to Γ has intersection form (see for instance, [HNK])

$$A(\Gamma) = (\alpha_{ij}) \text{ with}$$

$$\alpha_{ij} = \begin{cases} 1 & \text{if } i \neq j \text{ and } v_i \text{ and } v_j \text{ are connected in } \Gamma, \\ b_i & \text{if } i = j, \\ 0 & \text{otherwise.} \end{cases}$$

We call Γ *positive definite*, *negative definite*, or *even* according to whether $A(\Gamma)$ has these properties.

Note that if Γ is positive definite, then blowing up or down +1 vertices does not change this property; similarly, for negative definiteness and -1 vertices.

Theorem 5.2. Let M *be an arbitrary Seifert manifold. Then* M *can be written as* $M \cong \partial P(\Gamma)$ *as in Theorem* 5.1 *with* Γ *definite if and only if* $e(M) \neq 0$. *In this case* Γ *is positive definite or negative definite according as* $e(M) > 0$ *or* $e(M) < 0$, *and* Γ *is unique after blowing down all* +1 *vertices, respectively all* -1 *vertices, which can be blown down in von Randow's sense.*

Proof. Let Γ be a graph as in Theorem 5.1. Then a simple induction shows $A(\Gamma)$ can be diagonalized as

$$\mathrm{diag}(e(M), c_{11}, \ldots, c_{1,s_1}, c_{22}, \ldots, c_{2,s_2}, \ldots, c_{m,s_m})$$

with

$$c_{ij} = [b_{ij}, b_{i,j+1}, \ldots, b_{i,s_i}] .$$

Thus Γ can be positive definite if and only if $e(M) > 0$, and similarly for negative definite. Assume now $e(M) > 0$, by reversing the orientation of M if necessary. Normalize the Seifert invariant of M to satisfy $0 < \beta_i < \alpha_i$ for $i = 1, \ldots, m$. Then α_i/β_i can be uniquely expanded as a continued fraction

$$\alpha_i/\beta_i = [b_{i1}, \ldots, b_{is}] \text{ with } b_{ij} \geqq 2 .$$

A simple induction then shows $c_{ij} > 0$ for all i, j, so $A(\Gamma)$ is positive definite.

Conversely, suppose $A(\Gamma)$ is positive definite, and by blowing down if necessary, assume no b_{ij} equals 1. Then $b_{ij} \geqq 2$ for all i, j (since $b_{ij} > 0$ by positive definiteness). This forces $0 < \beta_i < \alpha_i$ for the Seifert invariants, so the Seifert invariants are in normalized form, hence unique, and the b_{ij} are then uniquely determined by the above comments.

Corollary 5.3. Let M be a Seifert manifold with $e(M) \neq 0$. Then by reversing orientation if necessary we can assume $e(M) < 0$, and M is then the link of an isolated surface singularity with C^*-action, and the "canonical negative definite graph Γ for M" given by the above theorem is the dual graph of a resolution of this singularity.

Proof. One can do plumbing holomorphically to obtain $P(\Gamma)$ as a complex manifold with holomorphic C^*-action and then apply Grauert [G] to blow down the central configuration of curves in $P(\Gamma)$. That one can blow down equivariantly follows by functoriality of Grauert's theorem. Since the complex structure one puts on $P(\Gamma)$ is in general far from unique, one, of course, gets a whole family of possible singularities. One can also prove this corollary directly via the injective holomorphic C^*-actions of Conner and Raymond [CR1], by showing that they can be compactified by a single singular point to give a complex affine variety, if the Euler number is negative.

Remark. The resolution given by the above corollary is the minimal "good" resolution of the singularity. In some cases, one can blow down further. For example, Theorem 2.1 shows that the link $\Sigma(2,11,19)$ of the singularity of $V(2,11,19) = \{z \in C^3 / z_1^2 + z_2^{11} + z_3^{19} = 0\}$ is the Seifert manifold $M(0;(1,1),(2,-1),(11,-2),(19,-6))$ so Corollary 5.3 gives the graph

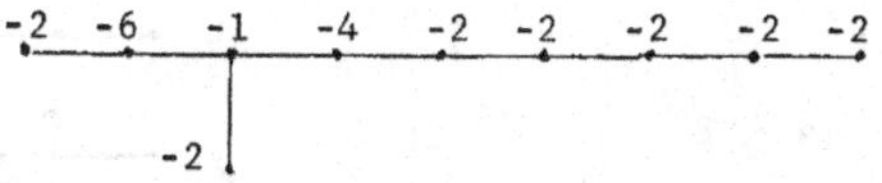

as the dual graph of the minimal good resolution. Blowing down (in the sense of complex manifolds; we cannot do it in the sense of plumbing graphs, since the (-1)-vertex has three neighbors) can be done twice, giving the result

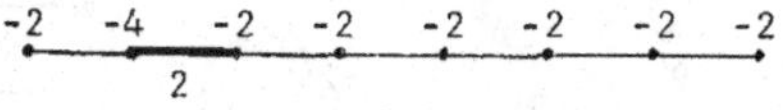

where the heavy line means a tangency of intersection number 2 between the corresponding curves of the resolution. Since $\Sigma(2,11,19)$ is the boundary of a regular neighborhood of the corresponding configuration of curves, this shows that $\Sigma(2,11,19)$ bounds a simply connected four manifold Y with negative definite intersection form of signature -8 (which must hence be equivalent to the standard E_8 form, by the classification of such forms, but this can easily be seen directly).

It is of interest to know which homology spheres bound simply connected manifolds with even definite intersection forms. For Seifert homology spheres, the minimal resolution of the corresponding singularity will sometimes provide a positive answer. For example, the minimal good resolution for $\Sigma(2,4k-1,8k-3)$, or what is the same, the canonical plumbing diagram, has even form of signature $-8k$. It is, in fact, the bilinear form commonly denoted Γ_{8k}. $\Sigma(2,8k-5,12k-7)$ is another example giving even forms of signature -8k .

The following table gives all $\Sigma(p,q,r)$ with $p < q < r$ pairwise coprime, $p = 2$ and $q < 20$, or $p \leq 5$ and $q \leq 10$, for which the minimal resolution gives what we want. Omitted weights are -2 . Double lines represent curves of the resolution intersecting tangentially with intersection number 2 . Triangles represent three curves intersecting transversally in one point.

TABLE

(p,q,r)	signature	resolution graph
2 , 3 , 5	- 8	
2 , 7 , 13	-16	
2 , 11 , 17	-16	
2 , 11 , 19	- 8	
2 , 11 , 21	-24	
2 , 13 , 21	- 8	
2 , 15 , 29	-32	
2 , 19 , 29	-24	
2 , 19 , 37	-40	
3 , 4 , 7	- 8	
3 , 5 , 13	- 8	
3 , 7 , 17	- 8	
4 , 5 , 19	-24	
4 , 7 , 9	-16	
4 , 7 , 27	-32	
5 , 7 , 27	- 8	
5 , 9 , 13	- 8	
5 , 9 , 31	- 8	
5 , 9 , 43	-24	

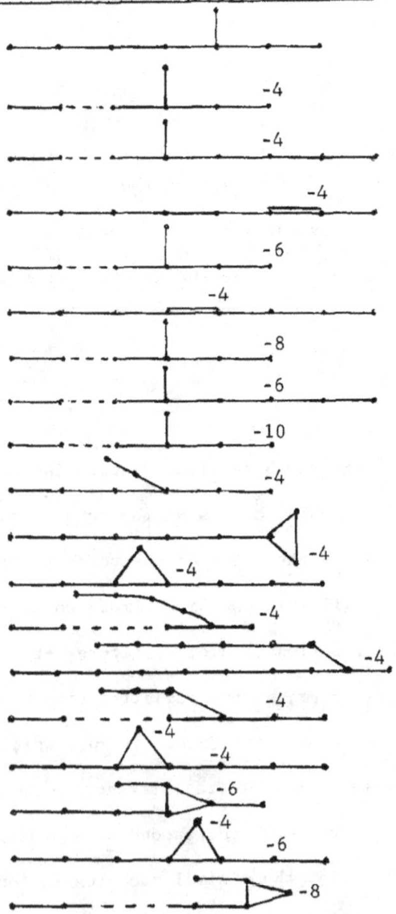

One can simplify the algorithm provided by Theorems 2.1 and 5.1 for finding a plumbing diagram for $\Sigma(a_1,\dots,a_n)$ by observing that if

$$[b_1,\dots,b_s] = p/q\ ,\quad b_i \geq 2$$

then

$$[b_s,b_{s-1},\dots,b_1] = p/q'$$

with $0 < q \leq p$, $0 < q' \leq p$, $qq' \equiv 1 \pmod p$. We thus get the

Algorithm 5.4. If $a_1, \dots, a_n$ *are pairwise coprime integers with* $a_j \geq 2$, *define* $c_1, \dots, c_n$ *by*

$$c_j \equiv -a_1 \dots \hat{a}_j \dots a_n \pmod{a_j}\ ,\quad 0 < c_j < a_j\ .$$

Expand a_j/c_j *as a continued fraction*

$$a_j/c_j = [d_{j1},d_{j2},\dots,d_{j,s_j}]\ ,\quad d_{ji} \geq 2\ .$$

Then

$\Gamma =$ 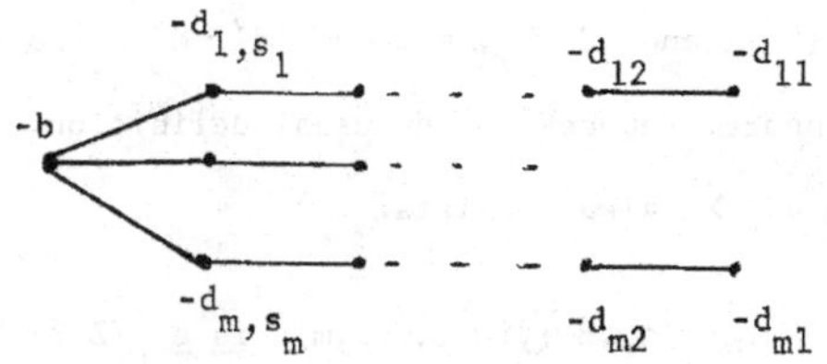

is the canonical (in the sense of 5.2 and 5.3) plumbing graph for $\Sigma(a_1,\dots,a_n)$, *where* b *is best determined in practice by estimating it via the equation*

$$b = \sum_j 1/[d_{j,s_j},\dots,d_{j2},d_{j1}] + 1/a_1 \dots a_n$$

We leave the proof to the reader as well as the generalization to non-coprime a_i.

6. μ-Invariant for Seifert Manifolds

We first give a slightly generalized version of the usual definition of μ-invariant. If X^4 is any oriented four-manifold, a class $d \in H_2(X;Z)$ will be called a spherical integral Wu class if it satisfies:

a) d can be represented by a smoothly embedded sphere;

b) The mod 2 reduction w of d satisfies $x \cdot x = x \cdot w$ for all $x \in H_2(X;Z/2)$ (dot is intersection number).

Theorem 6.1. *Let M^3 be a $(Z/2)$-homology sphere. Let X^4 be a 4-manifold with a spherical integral Wu class d, such that $\partial X^4 = M^3$. Then*

$$\mu(M) = \text{sign } X - d \cdot d \pmod{16}$$

is an invariant of M.

Proof. If X', d' is another such pair and $Y = X \cup (-X')$ pasted along the boundary M, then the Meyer Vietoris sequence shows $H_2(Y;Z/2) \cong H_2(X;Z/2) \oplus H_2(X';Z/2)$ and $d_Y = d + d'$ is a spherical integral Wu class for Y. By a theorem of Kervaire and Milnor [KM], $\text{sign}(Y) - d_Y \cdot d_Y$ is divisible by 16. Since $\text{sign}(Y) = \text{sign}(X) - \text{sign}(X')$ and $d_Y \cdot d_Y = d \cdot d - d' \cdot d'$, the theorem follows. Note that if $d = 0$ the theorem reduces to the usual definition of μ-invariant, and as is well known, such an X always exists.

Theorem 6.2. *$M = M(g\,;(\alpha_i, \beta_i), i = 1,\dots,m)$ is a $(Z/2)$-homology sphere if and only if $g = 0$ and either:*

(i) all the α_i are odd and $\Sigma \beta_i$ is odd; or

(ii) exactly one of the α_i is even, say α_1.

In these cases the μ-invariant is given respectively by

(i) $\mu(M) = \Sigma_{i=1}^{m} (c(\alpha_i, \beta_i) + \text{sign } \beta_i) + \text{sign } e(M) \pmod{16}$

(ii) $\mu(M) = \Sigma_{i=1}^{m} c(\alpha_i - \beta_i, \alpha_i) + \text{sign } e(M) \pmod{16}$

where $c(p,q)$ is the function introduced in [N1] described below, $e(M) = -\Sigma \beta_i/\alpha_i$, and in case (ii) we have chosen the Seifert invariant so that $(\alpha_i - \beta_i)$ is odd for all i (possible, by replacing β_i by $\beta_i \pm \alpha_i$ if necessary for each $i > 1$, and then adjusting β_1 so $e(M)$ is unchanged).

Here $c(p,q)$ is defined for coprime integer pairs (p,q) with p odd. It is uniquely determined by the recursions

$$c(p,\pm 1) = 0$$
$$c(p,-q) = c(-p,q) = -c(p,q)$$
$$c(p,p+q) = c(p,q) + \operatorname{sign}(q(p+q))$$
$$c(p\pm 2q,q) = c(p,q)\,.$$

These recursion formulae, in fact, give the fastest computation of $c(p,q)$ in practice, but various other descriptions of $c(p,q)$ are known. Before proving the above theorem, we describe some of them.

<u>Proposition 6.3</u>. <u>If</u> $p\,,q > 0$ <u>then</u>

(i) $c(p,q) = \mu(L(q,p)) \pmod{16}$, q <u>odd</u>;

(ii) $c(p,q) = \alpha(L(q,p),T)$, <u>where</u> α <u>is the Browder Livesay invariant and</u> T <u>is the involution on</u> $L(q,p)$ <u>with orbit space</u> $L(2q,p)$;

(iii) $c(p,q) = \Sigma(-1)^i \ \#\{0<k<q \mid i<kp/q<i+1\}$,

$$= -\frac{1}{q}\sum_{\eta^q=-1}\frac{(\eta+1)(\eta^p+1)}{(\eta-1)(\eta^p-1)}\ ,$$

$$= q-1-4N_{p,q}\ ,\quad q \ \text{odd}\ ,$$

where $N_{p,q} = \#\{1 \le i \le \frac{q-1}{2} \,\big|\, \frac{q-1}{2} < pi < q \pmod{q}\}$.

<u>Proof</u>. (i) is the special case of Theorem 6.2 with $M = M(0\,;(p,q)) = L(q,p)$ for q odd. Equation (ii) is a way $c(p,q)$ originally came up in [N1]. This function was renamed $t(q;p)$ and generalized by Hirzebruch and Zagier ([H1], [HZ], especially pp. 245-246) and (iii) is a selection of the many formulae for $t(q;p)$ given there. The last formula is especially interesting, since by Gauss, $(-1)^{N_{p,q}} = (\frac{p}{q})$ is the quadratic residue symbol, so the last formula implies

$$c(p,q) \equiv 1 - 2(\tfrac{p}{q}) + q \pmod{8}\ ,\quad q \ \text{odd}\ .$$

This was first observed by Hirzebruch [HNK].

We have appended, at the end of this section 6, a table from [N-1] for $c(p,q)$. All values for $p \leq 27$ and $q \leq 26$ are given.

Proof of Theorem 6.2. Firstly, as in the proof of Theorem 4.1, one sees that $M = M(g\,;(\alpha_i,\beta_i))$ is a $(Z/2)$-homology sphere if and only if $g = 0$ and $\Sigma\beta_i\alpha_1 \ldots \hat{\alpha}_i \ldots \alpha_m$ is odd. This implies the first statement of the theorem.

To see the formulae for $\mu(M)$, we first consider case (ii). That is, we assume $M = M(0\,;(\alpha_i,\beta_i), i = 1,\ldots,m)$ with α_1 even, α_i odd for $i > 1$. As already remarked, we can also assume β_i is even for $i > 1$, by replacing β_i by $\beta_i \pm \alpha_i$ if necessary.

Lemma 6.3. *If p and q are coprime integers, then p/q has a continued fraction expansion*

$$p/q = [b_1,b_2,\ldots,b_s]$$

(see Theorem 5.1) with each b_i even if and only if exactly one of p and q is even. There is then a unique such expansion satisfying in addition: $|b_i| \geq 2$ for $i > 1$.

The proof of this lemma is an easy induction which we omit.

Applying this lemma and Theorem 5.1, we can express $M^3 = \partial P(\Gamma)$, where Γ is a weighted tree

$\Gamma =$ 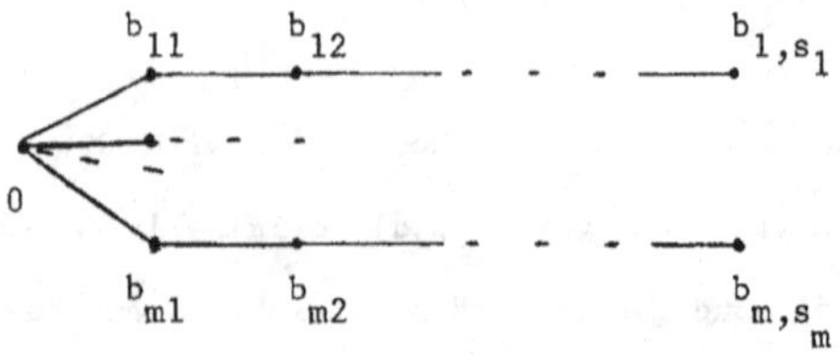

with all weights even, $|b_{ij}| \geq 2$ for $j > 1$, and $\alpha_i/\beta_i = [b_{i1},\ldots,b_{i,s_i}]$ for each i. If we take $X = P(\Gamma)$, so $M = \partial X$, then the definition of $\mu(M)$ reduces to $\mu(M) = \text{sign}\,(X) \pmod{16}$. Using the diagonalization of the intersection matrix $A(\Gamma)$ of $X = P(\Gamma)$ described in the proof of Theorem 5.2, we see

$$\text{sign}\, X = \sum \mu(\alpha_i,\beta_i) + \text{sign}\, e(M)$$

where $\mu(\alpha_i,\beta_i) = \#\{j|1<j\le s_i, b_{ij}>0\} - \#\{j|1<j\le s_i, b_{ij}<0\} + \operatorname{sign}(\alpha_i/\beta_i)$. The following recursion formulae follow directly from this definition of $\mu(\alpha,\beta)$:

$$\mu(\alpha,\beta) \text{ is defined if } \alpha+\beta \text{ is odd and } \gcd(\alpha,\beta) = 1;$$

$$\mu(\alpha,\beta) = -\mu(\alpha,-\beta);$$

$$\mu(2b\alpha-\beta,\alpha) = \mu(\alpha,\beta) + \operatorname{sign} b \quad \text{if } |\alpha| > |\beta|;$$

$$\mu(\beta,\alpha) = -\mu(\alpha,\beta) + \operatorname{sign}(\beta/\alpha) \quad \text{if } |\alpha| > |\beta|.$$

If we define $c'(p,q) = \mu(q,q-p)$, then it is easy to deduce that c' satisfies the recursion formulae defining c, so $c'(p,q) = c(p,q)$. Thus $\mu(q,q-p) = c(p,q)$, so $\mu(\alpha,\beta) = c(\alpha-\beta,\alpha)$, completing the proof of the formula for case (ii).

The proof in case (i) can be done similarly, although in this case M cannot be written as $\partial P\Gamma$ where Γ has only even weights, which complicates this approach slightly. A more interesting proof uses the following theorem.

<u>Theorem 6.4</u>. <u>Let</u> M^3 <u>be a</u> $(Z/2)$-<u>homology sphere and</u> $T: M \to M$ <u>a free orientation preserving involution</u>. <u>Suppose</u> $M = \partial X^4$ <u>and</u>

a) T <u>extends to</u> $T': X \to X$;

b) T' <u>acts trivially on</u> $H_2(X,\mathbb{Q})$ <u>and</u> $H_2(X;Z/2)$;

c) <u>the</u> 2-<u>dimensional part</u> F <u>of</u> $\operatorname{Fix}(T')$ <u>is oriented and is homologous to a smoothly embedded</u> 2-sphere.

<u>Then</u>

$$\alpha(M,T) \equiv \mu(M) \pmod{16}$$

<u>where</u> $\alpha(M,T)$ <u>is the Browder Livesay invariant</u>.

<u>Proof</u>. By Hirzebruch [H2] (see also [HJ] and [AS])

$$\alpha(M,T) = \operatorname{sign}(X,T') - [F]\cdot[F]$$

where $[F] \in H_2(X;Z)$ is the represented homology class. Since T' acts trivially on $H_2(X;\mathbb{Q})$, we have $\operatorname{sign}(X,T') = \operatorname{sign} X$. We must thus only show that $[F]_2 \in H_2(X;Z/2)$ satisfies $[F]_2 \cdot x = x \cdot x$ for all $x \in H_2(X;Z/2)$. But $x = T'x$ by assumption (b), and $[F]_2 \cdot x = x \cdot T'x$, since if we represent x by a cycle C, then the intersection points of C and $T'C$ pair off under T', and thus contribute nothing to $x \cdot T'x$, unless they lie in $F \cap C$.

To apply this theorem to case (i) in Theorem 6.2, observe that in this case the involution T contained in the S^1-action on M is a free involution, and if we write $X = \partial P(T)$ as in Theorem 5.1, then T extends to $T' : X \to X$ since the whole S^1-action extends. Condition (b) is satisfied since T' is homotopic to the identity, and (c) is satisfied since $\mathrm{Fix}(T')$ is a union of disjoint spheres (the zero-sections of some of the bundles being plumbed). Thus $\mu(M) \equiv \alpha(M,T) \pmod{16}$. But $\alpha(M,T)$ was computed in [N] as

$$\alpha(M,T) = \sum (c(\alpha_i,\beta_i) + \mathrm{sign}\,\beta_i) + \mathrm{sign}\, e(M)$$

whenever $M = M(g,(\alpha_i,\beta_i))$ with all the α_i odd, so the proof is completed.

If M is a Z-homology sphere, other formulae for $\mu(M)$ are known, in view of the fact that $M \cong \Sigma^3(\alpha_1,\ldots,\alpha_n)$ up to orientation.

<u>Theorem 6.5</u>. $M = \Sigma^3(\alpha_1,\ldots,\alpha_n)$ <u>with</u> α_i <u>pairwise coprime embeds in</u> S^5 <u>as a fibered knot, the signature of whose fiber</u> V <u>is</u>

$$\mathrm{sign}(V) = \sum_{\substack{1 \le j < 2\alpha \\ j \text{ odd}}} \mathrm{res}_{\pi i j/\alpha}\Big((\tanh \alpha z)^{n-2} \coth z \prod_{k=1}^{n} \coth \frac{\alpha z}{\alpha_k}\Big) ,$$

$$= t(\alpha_1,\alpha_2,\alpha_3) \text{ of [HZ] if } n = 3 .$$

<u>In particular</u>, $\mu(M) = \mathrm{sign}(V) \pmod{16}$.

<u>Proof</u>. If $\Sigma^5(\alpha_1,\ldots,\alpha_n)$ is defined just like $\Sigma^3(\alpha_1,\ldots,\alpha_n)$ but using $(n-3)$ instead of $(n-2)$ equations, then by Hamm [Ha], there is a "Milnor fibration" of the complement of Σ^3 in Σ^5, whose fiber V has the above signature (see also Hirzebruch [H3]). Furthermore, if $\alpha_1, \ldots, \alpha_n$ are pairwise coprime, then Σ^5 is a homotopy sphere, so $\Sigma^5 \cong S^5$. V is stably parallelizable, so its intersection form is even, so $\mu(M) = \mathrm{sign}(V) \pmod{16}$ if $\mu(M)$ is defined (e.g., α_i pair-

wise coprime).

For $n = 3$ the function in the above theorem was denoted $t(\alpha_1, \alpha_2, \alpha_3)$ and studied and tabulated by Hirzebruch and Zagier ([HZ], table on page 118).

TABLE OF $c(p,q)$

q\p	1	3	5	7	9	11	13	15	17	19	21	23	25	27
1	0	0	0	0	0	0	0	0	0	0	0	0	0	0
2	1	-1	1	-1	1	-1	1	-1	1	-1	1	-1	1	-1
3	2		-2	2		-2	2		-2	2		-2	2	
4	3	1	-1	-3	3	1	-1	-3	3	1	-1	-3	3	1
5	4	0		0	-4	4	0		0	-4	4	0		0
6	5		1	-1		-5	5		1	-1		-5	5	
7	6	2	2		-2	-2	-6	6	2	2		-2	-2	-6
8	7	1	-1	1	-1	1	-1	-7	7	1	-1	1	-1	1
9	8		0	0		0	0		-8	8		0	0	
10	9	3		3	1	-1	-3		-3	-9	9	3		3
11	10	2	2	-2	2		-2	2	-2	-2	-10	10	2	2
12	11		3	1		1	-1		-1	-3		-11	11	
13	12	4	0	0	4	0		0	-4	0	0	-4	-12	12
14	13	3	1		-3	-1	1	-1	1	3		-1	-3	-13
15	14			2		2	2		-2	-2		-2		
16	15	5	3	1	-1	5	3	1	-1	-3	-5	1	-1	-3
17	16	4	4	4	0	-4	0	0		0	0	4	0	-4
18	17		1	-1		-1	1		1	-1		-1	1	
19	18	6	2	2	2	2	6	-2	2		-2	2	-6	-2
20	19	5		1	3	1	-5		-1	1	-1	1		5
21	20		4			0	0		4	0		0	-4	
22	21	7	5	3	5		1	7	1	3	1	-1	-3	-1
23	22	6	2	2	-2	2	-2	-6	2	2	2		-2	-2
24	23		3	5		1	-1		3	-3		1	-1	
25	24	8		0	0	0	0		8	0	0	0		0
26	25	7	5	3	1	3		3	-7	3	5	-1	1	-1

7. μ-Invariant for Plumbed Manifolds

Theorem 6.1 enables us to give an algorithm to compute $\mu(M)$ for an arbitrary $(Z/2)$-homology sphere obtained by plumbing. Note that a necessary condition that $M = \partial P(\Gamma)$ be a $(Z/2)$-sphere is that Γ be a tree and all the genus weights vanish.

Theorem 7.1.

(i) *Let* $M = \partial P(\Gamma)$ *be the result of plumbing bundles over* S^2 *according to a tree* Γ. *Then* M *is a* $(Z/2)$-*sphere if and only if* Γ *can be reduced to a collection of isolated points with odd weights by a sequence of moves of type* 1 *and* 2 *below.* M *is not a* $(Z/2)$-*sphere if and only if* Γ *can be so reduced to a collection of isolated points with at least one even weight.*

Let

$\Gamma =$ 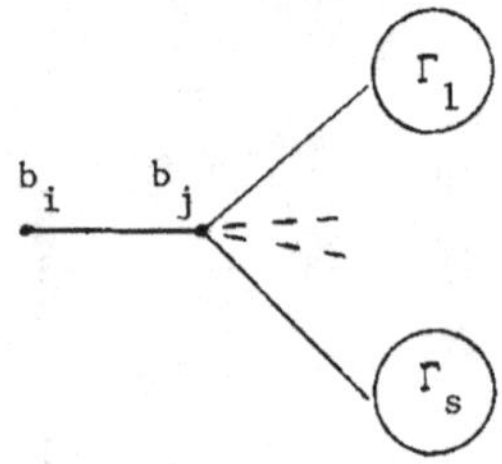

where b_i *and* b_j *are the weights of vertices* i *and* j.

Move 1. *If* b_i *is even, replace* Γ *by the disjoint union* Γ' *of* $\Gamma_1, \ldots, \Gamma_s$.

Move 2. *If* b_i *is odd, replace* Γ *by*

$\Gamma' =$ 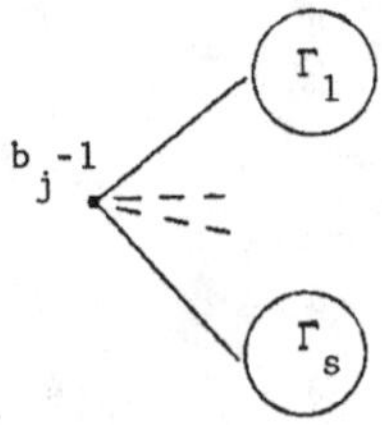

(ii) *If* $M = \partial P(\Gamma)$ *is a* $(Z/2)$-*sphere, define a subset* $S(\Gamma)$ *of the vertices of* Γ *inductively as follows:*

a) *If* Γ_0 *is a set of isolated points with odd weights, put* $S(\Gamma_0)$ *equal to the set of all these points.*

b) *If* $S(\Gamma')$ *is known and* Γ *reduces to* Γ' *by Move 1 above, put*

$$S(\Gamma) = S(\Gamma') \cup \{i\}\ ,$$
$$= S(\Gamma')\ ,$$

according as the number of points in $S(\Gamma')$ *adjacent to vertex* j *is congruent to* $b_j - 1$ *or* b_j *modulo* 2.

c) *If* Γ *reduces to* Γ' *by Move* 2, *put*

$$S(\Gamma) = S(\Gamma') \cup \{i\} \quad \text{if} \quad j \notin S(\Gamma')\ ,$$
$$= S(\Gamma') \quad \text{if} \quad j \in S(\Gamma')\ .$$

Then

$$\mu(M) = \operatorname{sign} A(\Gamma) - \sum_{i \in S(\Gamma)} b_i \quad (\text{mod } 16),$$

where $A(\Gamma)$ *is the matrix of the graph.*

Proof. $H_1(\partial P(\Gamma)) \cong \operatorname{Cok}(A(\Gamma))$, so $M = \partial P(\Gamma)$ is a $(Z/2)$-sphere if and only if $\det A(\Gamma)$ is odd. But it is easily verified that Moves 1 and 2 do not change $\det A(\Gamma)$ (mod 2), so the first part of the theorem follows. For the second part, let $X = P(\Gamma)$ and let $\{e_i \mid i$ a vertex of $\Gamma\}$ be the standard basis of $H_2(P(\Gamma);Z)$ represented by the zero-sections of the plumbed bundles. Then a simple induction shows that $d = \sum_{i \in S(\Gamma)} e_i$ is a spherical integral Wu class for X and that $d \cdot d = \sum_{i \in S(\Gamma)} b_i$. Since $\operatorname{sign} X = \operatorname{sign} A(\Gamma)$, the theorem follows.

Problem 7.2. If M *is a* $(Z/2)$-*homology* 3-*sphere with a free orientation preserving involution, is it true that* $\mu(M) = \alpha(M,T)$ (mod 16) ?

The answer is "yes" for Seifert Z-homology spheres, and more generally for Seifert $(Z/2)$-spheres $M(0;\alpha_i,\beta_i))$ with pairwise distinct Seifert pairs $(\alpha_i,\beta_i \bmod \alpha_i)$. In these cases we shall show in a later paper that any free involution must be in the S^1-action, putting us into the situation of the proof of part (i) of Theorem 6.2.

8. Orientation Reversing Maps

It turns out that many properties of a Seifert manifold are determined by its Euler number $e(M)$.

Theorem 8.1. *If $e(M) \neq 0$, then $\alpha_1 \cdots \alpha_m |e(M)|$ = order of torsion of $H_1(M;Z)$. If $e(M) = 0$, then M fibers equivariantly over the circle. If M is not a principal circle bundle over a torus, then M fibers over the circle if and only if the fibering is equivariant.*

This is due to Orlik, Vogt and Zieschang [OVZ] and Orlik and Raymond in certain exceptional cases. The fibering, if it exists, is far from unique. These fiberings have been constructed explicitly from the viewpoint of homologically injective actions by Conner and Raymond [CR2]. That these fiberings are S^1-equivariant follows most easily from this viewpoint. The principal circle bundles over the torus are the only Seifert fiberings which fiber over the circle but fail to fiber equivariantly. (All but the 3-torus has $e(M) \neq 0$.) The S^1-equivariant fiberings are also constructed explicitly from a plumbing viewpoint by Neumann in [N3].

The present investigation arose from the next

Theorem 8.2. *If the Seifert manifold M is not a lens space, then the following statements are equivalent:*

(i) *M admits a free orientation reversing involution*

(ii) *M admits an orientation reversing involution*

(iii) *M admits an orientation reversing homeomorphism*

(iv) *M admits an orientation reversing self-homotopy equivalence*

(v) *M fibers over S^1 and admits an orientation reversing free involution which commutes with the S^1-action.*

Proof. Clearly (v)$\Longrightarrow$(i)$\Longrightarrow$(ii)$\Longrightarrow$(iii)$\Longrightarrow$(iv). We show (iv)$\Longrightarrow$(v). We suppose first that M has infinite fundamental group. We assume M is not the 3-torus.

Then by [W] if M is sufficiently large or by [OVZ] or [CR3], in general, any homotopy equivalence $M \to -M$ is homotopic to a fiber preserving homeomorphism, so $e(M) = e(-M) = -e(M)$. Hence, $e(M) = 0$. Therefore, M fibers over S^1 equivariantly. In fact, since M has Seifert invariants $(g ; (\alpha_i, \beta_i))$ then $-M$ has Seifert invariants $(g;(\alpha_i,-\beta_i))$. This easily yields that the Seifert invariants of M must be expressible as

$$M = (g ; (2,b_1),\ldots,(2,b_s),(\alpha_i,\beta_i),(\alpha_i,-\beta_i))$$

for some $s, k \geq 0$, $\alpha_i > 2$, $i = 1, \ldots, k$. Now $e(M) = 0$ implies $\frac{1}{2}\Sigma b_i = 0$ and since the b_i are odd, this implies s <u>is even</u>. Thus, <u>the Seifert invariants for</u> M <u>are equivalent to</u>

(vi) $$(g ; (\alpha_i,\beta_i),(\alpha_i,-\beta_i), i = 1,\ldots,\ell).$$

Therefore, M is the orientation double covering fixed-point free S^1-manifold by Seifert [S;p. 198]. This completes the proof if $\pi_1(M)$ is infinite.

Our attack for the finite fundamental groups must be different. Each Seifert manifold with finite non-abelian fundamental group appears as S^3/G where G is a finite subgroup of SO(4) which acts freely on S^3, that is, a spherical space form. The 2-Sylow subgroups of these manifolds are either cyclic of order at least 4 or a generalized quaternionic group. The 2-Sylow subgroups are all conjugate and in the cyclic case, there is a unique subgroup of order 4. The generalized quaternionic groups have a characteristic subgroup of order 4, namely the second term of the upper central series. The quaternion group itself has a unique conjugacy class of elements of order 4. In any case, we may pass to the unique covering space corresponding to the subgroup of order 4 since this is determined up to conjugacy.

This must be a lens space L(4,1) or L(4,3). Whether it is L(4,1) or L(4,3) will be determined by the orientation of M. Now any self homotopy equivalence f of M must preserve the unique conjugacy class of our subgroup of order 4. Hence, f may be lifted to a self homotopy equivalence $\tilde{f}: L(4,a) \to L(4,a)$.

If f reverses orientation, then $\tilde{f}$ must do the same. But, $L(4,a)$, $a = 1$, or 3 admits no orientation reversing self-homotopy equivalence since -1 is not a square modulo 4. Hence, M could not possess an orientation reversing self-homotopy equivalence. This completes the proof of Theorem 8.2.

Remarks 8.3. It has been recently shown by C.B. Thomas, [T] that if M is a closed 3-manifold whose universal covering is the 3-sphere, then $\pi_1(M)$ must be one of the fundamental groups of Seifert manifolds with finite fundamental group. Since it is also known that any free $Z/4$ action on the 3-sphere [Ri] yields a lens space, we may conclude that the argument above also shows that such manifolds admit no orientation reversing self-homotopy equivalences. Of course, no examples of closed 3-manifolds with finite fundamental group which fail to be Seifert manifolds are known at this time and so this remark may be redundant.

Our arguments extend to the other types of oriented Seifert 3-manifolds which have not been considered elsewhere in this paper. We describe this situation now.

8.4. The Closed Case

We assume that M is a closed oriented Seifert 3-manifold with a non-orientable decomposition space. The fibering mapping is not the orbit mapping of an S^1-action and its type is distinct from the Seifert manifolds considered elsewhere in this paper. With a few exceptions, none of these manifolds support an S^1-action. They do support "local $SO(2)$- actions". The Seifert invariants are written

$$(On\ k\ ;\ (\alpha_i, \beta_i))$$

where the On refers to orientable total space and non-orientable base. They are exactly similar to the invariants for oriented Seifert fiberings with orientable decomposition space except that k represents the non-orientable genus of the decomposition space, and so, $k \geq 1$. Just as before the unnormalized representation is not unique.

We first observe that there is a double covering M' of M which is an oriented Seifert manifold with orientable decomposition space and whose unnormalized invariants are

$$(g=k-1\,;\,(\alpha_1,\beta_1),(\alpha_1,\beta_1),\ldots,(\alpha_s,\beta_s),(\alpha_s,\beta_s))\,.$$

One can easily deduce from this that the naturality properties of $e(M) = -\Sigma\beta_i/\alpha_i$ extend to the case of non-orientable decomposition space as long as the total space M is kept orientable. But it is also easy to reduce considerations to the case of orientable decomposition space, which is the method we shall follow.

Assume, now, that if $k = 1$, then there are at least 2 singular fibers, if $k = 2$, then there is at least one singular fiber, and if $k = 1$, $s = 2$, then $\{(\alpha_1,\beta_1),(\alpha_2,\beta_2)\} \neq \{(2,1),(2,-1)\}$. We shall treat these presently avoided cases separately.

Under the hypothesis on the invariants, the element of the fundamental group represented by an ordinary fiber generates an infinite cyclic characteristic subgroup of $\pi_1(M)$ and $\pi_1(M')$ is the centralizer of this cyclic subgroup. It is easy to check that every automorphism of $\pi_1(M)$ induces an automorphism of the subgroup $\pi_1(M')$. Consequently, we may lift any self-homotopy equivalence h on M to a self-homotopy equivalence h' on M'. h will be orientation reversing if and only if h' is orientation reversing. Therefore, we know that M' must be of the type exhibited in (vi) of Theorem 8.2. Consequently, the Seifert invariants of M must be

$$(\mathrm{Onk}\,;\,(\alpha_i,\beta_i),(\alpha_i,-\beta_i))\,.$$

We now wish to show that each such M actually admits an orientation reversing involution. From p. 198 of [S], observe that as long as the non-orientable genus k of M is even, M is an orientable double covering of Seifert manifolds of type (N,n,II) and (N,n,III) using Seifert's terminology. For all $k > 1$, M is also an orientable double covering of certain non-orientable 3-manifolds closely related to the classical Seifert 3-manifolds. These manifolds described by Orlik and Raymond admit "local SO(2)-actions". Although they are not classical Seifert 3-manifolds, they would be considered as injective Seifert 3-manifolds from the point of view of Conner and Raymond. We see from the table on page 155 of [OR] that each M (type $n_2 = On$) is a double covering of a non-orientable local $SO(2)$-manifold, provided that $k > 1$. For $k = 1$, none of the tabulated double coverings will work. However, there exists an involution on $\mathbb{R}P_2$ so that an isolated point and orientation reversing circle appear as the fixed point set. This involution is embedded in the effective $SO(2)$-action on $\mathbb{R}P_2$. With this involution, one may define an involution on M where $k = 1$ and the Seifert invariants satisfy the necessary conditions for an orientation reversing homeomorphism. First, one removes the tubular neighborhoods of the singular fibers. The resulting circle bundle with structure group $O(2)$ is the restriction to the deleted $\mathbb{R}P_2$ of the associated sphere bundle $S(\theta \oplus 1)$ where θ is the line bundle $\det(T\mathbb{R}P_2)$ and 1 denotes the trivial line bundle. It makes sense, since the bundle has a section, to flip in the 1-direction. This carries the bundle over the region away from a Möbius band into itself by a rotation in D^2 and a flip in S^1. This can be extended to the deleted tubular neighborhoods. The involution has 2 isolated fixed points and so the orbit space is not a manifold. (No free involution presumably exists in case $k = 1$. The argument to check that no free involution exists is rather complicated and the details have not been checked.)

We turn now to the omitted cases. $M = (On1\ ; (2,1),(2,-1))$ has an involution as described above. $M = (On2\ ; (1,\beta))$ can also be regarded as a torus bundle over the circle with monodromy $\begin{pmatrix} -1 & 0 \\ -\beta & -1 \end{pmatrix}$, $\beta \in Z$. If $\beta \neq 0$, then the fundamental group of the torus fiber is a characteristic subgroup. The outer automorphism group of $\Pi_1(M)$ is calculated in Conner and Raymond [CR4;6.14]. It is readily seen from this calculation that if $\beta \neq 0$, M admits no orientation reversing self-homotopy equivalence. For $\beta = 0$, M can be identified with $\{g = 0,(2,1)(2,-1),(2,1),(2,-1)\}$ which does admit an orientation reversing free involution.

The remaining cases to treat are $M = M(On1\ ; (\alpha_1,\beta_1))$. If $(\alpha_1,\beta_1) = (1,0)$, then M is $\mathbb{R}P_3 \# \mathbb{R}P_3$ which certainly has an orientation reversing homeomorphism. If $\beta \neq 0$, then $M = M(On1;(\alpha,\beta))$ also has a Seifert fibering with orientable base as $M = (0;(2,1)(2,-1),(\alpha,\beta))$. If $\beta = \pm 1$ this is the lens space $\pm L(4\alpha, 2\alpha+1)$

and if $|\beta| > 1$ it is a nonabelian spherical space form. (This corrects a statement in [OR].) In either case it admits no orientation reversing equivalence.

We may now summarize our result for the closed case as follows:

Theorem 8.5. *The following are equivalent for* $M = M(Onk ; (\alpha_i, \beta_i))$ *not a lens space*:

(i) M *admits an orientation reversing self-homotopy equivalence*,

(ii) M *admits an orientation reversing homeomorphism*,

(iii) M *admits an orientation reversing involution*,

(iv) *The Seifert invariants may be written as*

$$(Onk ; (\alpha_j, \beta_j), (\alpha_j, -\beta_j)) .$$

Moreover, if $k > 1$, *the orientation reversing involution can be chosen to be free*.

8.6. M Compact But Not Closed

For this case we assume $\partial M \neq 0$. Let h denote the number of boundary components. Then the Seifert invariants for M are given by

$$((g,h) ; (\alpha_i, \beta_i))$$
$$(On(k,h) ; (\alpha_i, \beta_i)) ,$$

where we may assume that all $\alpha_i > 1$, $0 < \beta_i < \alpha_i$ and $i = 0, 1, 2, \ldots, m$. Similar to our procedure in 8.4, we assume that if $g = 0$, $h = 1$, then $m > 1$, and if $k = 1$, $h = 1$, then $m \neq 0$. We wish to consider only self-homotopy equivalences that preserve the "peripheral structure". Then, in order that $f : M \to M$ be such an orientation reversing self-homotopy equivalence, it must follow that

$$\{(\alpha_i, \beta_i)\} = \{(\alpha_1, \beta_1), (\alpha_1, \alpha_1 - \beta_1), \ldots, (\alpha_t, \beta_t), (\alpha_t, \alpha_t - \beta_t)\}$$

As before, involutions can be constructed on each of these manifolds. However, when the Euler characteristic of the decomposition space is odd, we cannot expect to find free involutions in general.

Added in Proof. The reason given in 8.2 for the quaternion group is incorrect. The result is still valid since the quaternion group can be embedded in SU(2). Consequently, the covering space associated to each subgroup of order 4 is the lens space L(4,3), or equivalently, the lens space L(4,1) if the opposite orientation of SU(2) is used. The rest of the argument proceeds as before.

BIBLIOGRAPHY

[CR1] Conner, P.E.;Raymond, F., Holomorphic Seifert fibering, Proceedings of the Second Conference on Compact Transformation Groups, Springer Lecture Notes, Vol. 299 (1972), 124-204.

[CR2] ______, Injective actions of toral groups, Topology 10 (1971), 283-296.

[CR3] ______, Deforming homotopy equivalences to homeomorphisms in aspherical manifolds, Bull. Amer. Math. Soc. 83 (1977), 36-85.

[CR4] ______, Manifolds with few periodic homeomorphisms, Proceedings of the Second Conference on Compact Transformation Groups, Springer Lecture Notes, Vol. 299 (1972), 1-75.

[G] Grauert, H., Über Modifikationen und exceptionelle analytische Raumformen, Math. Ann. 146 (1962), 331-368.

[H1] Hirzebruch, F., Free involutions on manifolds and some elementary number theory, Symposia Mathematica, Instituto Nazionale de Alta Matematica, Roma, Academic Press V (1971), 411-419.

[H2] ______, Involutionen auf Mannigfaltigkeiten, Proceedings of the Conference on Transformation Groups, Tulane (1967), Springer (1968), 148-166.

[H3] ______, Pontrjagin classes of rational homology manifolds and the signature of some affine hypersurfaces, Proceedings of Liverpool Singularities Symposium II (ed. C.T.C. Wall), Lecture Notes in Math. 209, Springer-Verlag (1971), 207-212.

[Ha] Hamm, H., Exotische Sphären als Umgebungsränder in speziellen komplexen Räumen, Math. Ann. 197 (1972), 44-56.

[HJ] Hirzebruch, F.;Jänich, K., Involutions and singularities, Algebraic Geometry, Papers presented at the Bombay Coll. (1968), Oxford University Press (1969), 219-240.

[HNK] Hirzebruch, F.;Neumann, W.D.;Koh,S.S., Differentiable manifolds and quadratic forms, Lecture Notes in Pure and Applied Mathematics Vol. 4, Marcel Dekker, New York, 1971.

[HZ] Hirzebruch, F.; Zagier, D., The Atiyah Singer theorem and elementary number theory, Math. Lecture Series 3, Publishor Perish Inc., Boston, Berkeley, 1974.

[K] Klein, F., Lectures on the Icosahedron and the solution of equations of the fifth degree, Dover, New York, 1956.

[KM] Kervaire,M.;Milnor, J., On 2-spheres in a 4-manifold, Proc. Nat. Acad. Sci. U.S.A. 49 (1961), 1651-1657.

[M] Milnor, J., On the 3-dimensional Brieskorn manifold M(p,q,r), Papers Dedicated to the Memory of R.H. Fox, Ann. of Math. Studies, Princeton University Press, 1975, No. 48, 175-225.

[N1] Neumann, W.D., S^1-actions and the α-invariant of their involutions, Bonner Math. Schriften 44, Bonn, 1970.

[N2] ______, Brieskorn complete intersections and automorphic forms, Invent.Math. 42 (1977), 285-293.

[N3] Neumann, W.D., Fibering graph manifolds over S^1, to appear.

[O] Orlik, P., Seifert manifolds, Springer Lecture Notes, Vol. 291 (1972).

[OR] Orlik, P.;Raymond, F., On 3-manifolds with local SO(2)-action, Quart. J. Math. Oxford (2) 20 (1969), 143-160.

[OW] Orlik, P.;Wagreich, P., Isolated singularities of algebraic surfaces with C^*-action, Ann. of Math. 93 (1971), 205-228.

[OVZ] Orlik, P.;Vogt, E.;Zieschang, H., Zur Topologie gefaserter dreidimensionaler Mannigfaltigkeiten, Topology 6 (1967), 49-64.

[Ran] Randell, R., The homology of generalized Brieskorn manifolds, Topology 14 (1975), 347-355.

[Ri] Rice, P.M., Free actions of Z_4 on S^3, Duke Math. J. 36 (1969), 749-751.

[RV] Raymond, F.;Vasquez, A.T., 3-manifolds whose universal coverings are Lie groups, to appear.

[S] Seifert, H., Topologie dreidimensionaler gefaserter Räume, Acta Math. 60 (1933), 147-238.

[T] Thomas, C.B., Homotopy classification of free actions by finite groups on S^3, to appear.

[vR] von Randow, R., Zur Topologie von dreidimensionalen Baummanigfaltigkeiten, Bonner Math. Schriften 14 (1962).

[W] Waldhausen, F., Eine Klasse von 3-dimensionalen Mannigfaltigkeiten, I and II, Invent. Math. 3 (1967), 308-333, 4 (1967), 87-117.

G SURGERY I - A SURVEY

by Ted Petrie

Rutgers University

Cat.

Organization of the Paper

This paper is organized as follows:

§0 Introduction - The main ideas and objectives are stated.

§1 Preliminaries - The main points of notation are recorded.

§2 Answers to the semi-classical questions - Statements of theorems pertaining to the question of §0 are given.

§3 Realization Theorems - Two general technical tools are given which deal with realizing invariants associated with components of fixed sets of a smooth G action.

§4 Realization Theorems and the theorems of §2 - A beginning is made of the outline of proofs of the theorems of §2. The first step applies the Realization Theorems.

§5 Combinatorics of smooth G manifolds - This section develops the combinatorial aspects of a smooth G manifold pertinent to the terms in the exact sequence 0.1.

§6 G surgery and $\tilde{K}_0(Z(G))$ - The projective class group appears forceably in the development of the theory encompassed by 0.1. This section outlines how and how it is treated.

§7 The terms in the exact sequence 0.1 - Here the terms in 0.1 are defined. Note in particular the definition of zero in $I^h(G,\lambda)$.

§8 Constructing elements in $N_G^h(Y,\lambda)$ - The first step in applying 0.1 starts with constructing elements of $N_G^h(Y,\lambda)$. The Realization Theorems of §3 are relevant here.

§9 Functorial properties of $I^c(G,\lambda)$ - This section is aimed at providing zero theorems for $I^c(G,\lambda)$ i.e. criteria for an element to vanish.

§10 Completing the proofs of §2 - The Realization Theorems have provided elements of $N_G^c(Y,\lambda)$ relevant to the theorems of §2. The exact sequence 0.1 together with the zero theorems of §9 are applied to show these elements map to zero in $I^c(G,\lambda)$.

§11 Motivating the definition of zero in $I^c(G,\lambda)$ - The essential points in the definition of zero in $I^c(G,\lambda)$ are reviewed and motivated.

§12 The obstruction theory - The entire theory encompassed by 0.1 is based on an obstruction theory which related the functors $I^c(\cdot,\lambda)$, $L_*(\cdot)$ and $\tilde{K}_0(\cdot)$. This relationship is essential for the properties of $I^c(\cdot,\lambda)$ in particular the exact sequence 0.1 and the properties of §9.

§0 Introduction

Let X and Y be smooth G manifolds where G is a finite group of order $|G|$. Let $f:X\to Y$ be a G map and define f to be an h equivalence if f is a homotopy equivalence. (This notion was called pseudo equivalence in [13] and [14].) The notion of an h equivalence singles out a distinguished family $\mathcal{P}$ of subgroups of G.

0.0 $$\mathcal{P} = \left\{P \subseteq G \,\middle|\, |P| \text{ is a prime power}\right\}.$$

Specifically if f is an h equivalence and $P \in \mathcal{P}$, then for each component α of $\pi_0(Y^P)$, f restricted to this component must induce an isomorphism $H_*(|\alpha|,R) \to H_*(f|\alpha|,R)$ where $|\alpha|$ is the space underlying α and R is a ring which depends on α and the set $\mathrm{Iso}(X) = \{G_x | x \in X\}$ of isotropy groups G_x of points x in X. This is an expression of Smith theory. It plays an essential role in the process of G surgery - a method for converting a G map into an h equivalence.

The aims of this paper are to survey the ideas of [14], [16], [17], [18] which deal with G surgery theory and to describe some of the problems to which the theory applies and outline their proofs in the context of the most important properties of the theory, in particular the exact sequence 0.1.

The process of G surgery to an h equivalence is summarized by the following exact sequence (see §7).

<u>Theorem 0.1</u>. $hS_G^h(Y,\lambda) \to N_G^h(Y,\lambda) \overset{d}{\underset{\to}{\subseteq}} I^h(G,\lambda)$ <u>is an exact sequence of sets</u>.

This generalizes the Wall exact sequence of [21].

In order to relate the requirements imposed by an h equivalence, chiefly derived from Smith theory, to the process of G surgery, we associate G posets (partially ordered G sets) $\Pi(Y) \supset \pi(Y)$ to a G space Y. By definition $\Pi(Y) = \coprod_{H \subset G} \pi_0(Y^H)$. The subspace $\pi(Y)$ is defined in terms of the set $\mathrm{Iso}(Y)$. When Y is a <u>smooth G manifold</u> $\pi(Y)$ determines $\Pi(Y)$ (§5) and provides a more natural record of the G invariants of Y for the process of G surgery which depends essentially on $\mathrm{Iso}(Y)$. The set $\Pi(Y)$ provides a more convenient device for expressing the requirements of an h equivalence. E.g. a G map $f : X \to Y$ induces G maps $\tilde{f} : \Pi(X) \to \Pi(Y)$ and $\hat{f} : \pi(X) \to \pi(Y)$ and if f is an h equivalence, $\tilde{f}$ induces a homeomorphism $\Pi(X,P) \to \Pi(Y,P)$ for $P \in \mathcal{P}$ if $\Pi(X,P) = \pi_0(X^P)$. This translates into the condition that $\hat{f}$ induces a homeomorphism $\pi(X,P) \to \pi(Y,P)$ for $P \in \mathcal{P}$.

By adding to $\pi(X)$ and $\pi(Y)$ some of the G invariants associated to components of fixed sets e.g. orientation classes for each component so that it is possible to discuss degree theory, we arrive at the notion of a G Poset pair λ which models the data $(\pi(X),\pi(Y),\hat{f},\mu) = \lambda$ where μ is a function supressed for now. It is on this data that the sets appearing in 0.1 are based, moreover, the data encompassed in λ provides the combinatorial format for the process of G surgery which for reasons of space cannot be explicitly discussed. See [15] §4 and §5.

The best justification and exemplification of the theory is provided by its applications. Here is a list of semi-classical questions to which the theory applies. Their treatment in the context of 0.1 is indicated in subsequent sections.

0.2 Which groups act on a closed 2n dimensional manifold with exactly one point fixed by the group?

Let G be a group which so acts. It is a consequence of the Atiyah-Bott fixed point theorem [2] that G cannot be cyclic of prime power order. See 2.7 and 2.8. This question is not so difficult. A considerably deeper question is

0.3 Which groups act on a closed homotopy sphere with exactly one fixed point?

This old question was generated by Montgomery-Samelson [10]. See Theorem 2.3 here and [13], [15] and [17].

0.4 <u>Which groups act on a closed homotopy sphere with exactly two fixed points and have distinct isotropy representations at the two fixed points?</u>

Let G be a group acting on a homotopy sphere Σ with this property. Atiyah and Bott show that $\mathrm{Iso}(\Sigma)$ contains at least three conjugacy classes of subgroups of G. See Theorems 2.4 and 2.5 here which demonstrate the necessity of this hypothesis.

0.5 <u>Given a group</u> G <u>and a ring</u> R, <u>does there exist a family</u> $\mathcal{K} = \mathcal{K}(G,R)$ <u>of subgroups of</u> G <u>such that</u> $G \notin \mathcal{K}$ <u>and dimension</u> Y^G <u>is a function of the dimensions of</u> Y^H <u>for</u> $H \in \mathcal{K}$ <u>whenever</u> Y <u>is a smooth closed homology</u> R <u>sphere</u>?

This question has an interesting history. Whenever Y is the unit sphere of a real G module V (real representation of G), then $\mathcal{K}$ may be taken to be the class of cyclic subgroups of G. This is an old theorem of E. Artin [9] and is related to the theorem that a representation is determined by its restrictions to all cyclic subgroups. The functional relation between $\dim Y^G$ and $\dim Y^H$ for H cyclic is called the Artin relation.

When G is an elementary abelian p group, the question has an affirmative answer due to Borel where $R = Z_p$. No further restrictions on Y except that it be a homology Z_p sphere are required. In this case $\mathcal{K}$ is the set of subgroups of G of index p plus the trivial group 1. In fact

$$\sum_{H \in \mathcal{K}} (\dim Y^H - \dim Y^G) = 0 \qquad [3]$$

Montgomery and Yang investigated 0.5 when $G = D_q$ is the dihedral group of order $2q$ where q is odd. The Artin relation for an action on Y <u>de-</u>

fined by a representation is

0.6 $$2 \dim Y^{Z_2} + \dim Y^{Z_q} = \dim Y^1 + 2 \dim Y^{D_q}$$

which explicitly represents $\dim Y^{D_q}$ as a function of $\dim Y^H$ for H cyclic. Montgomery and Yang showed that the left hand side of this equality always exceeds or equals the right hand side for an arbitrary smooth action on a homology sphere Y. They further conjectured equality [11].

Rothenberg (in a private communication) wondered whether it is possible to express $\dim Y^G$ as a function of $\dim Y^H$ for $H \neq G$ if it is required that Y^H be a homotopy sphere for all $H \subset G$.

The conjecture of Montgomery-Yang is false and the question of Rothenberg is negative. See Theorem 2.6.

§1 Preliminaries

Let Y be a compact G space and $\omega_G^0(Y)$ the equivariant 0th cohomotopy group of Y defined by

1.1 $$\omega_G^0(Y) = \varinjlim_{\Omega} [Y,F(\Omega)]^G$$

where the direct limit is over all real G modules (real orthogonal representations of G) Ω. $F(\Omega)$ is the space of self maps of the unit sphere $S(\Omega)$ of Ω. It is a G space via $(gf)(x) = gf(g^{-1}x)$ for $f \in F(\Omega)$, $x \in S(\Omega)$ and $g \in G$. The set of equivariant homotopy classes of maps of Y to $F(\Omega)$ is denoted by $[Y,F(\Omega)]^G$. Composition of maps in $F(\Omega)$ defines a multiplicative structure in $\omega_G^0(Y)$ and addition of maps (when $\dim \Omega^G > 0$) in $F(\Omega)$ defines an additive structure on $\omega_G^0(Y)$. It is a ring with unit 1 represented by mapping Y to the identity map of $S(\Omega)$.

The cohomology theory $\omega_G^0(\cdot)$ plays a special role in the theory of G surgery for two reasons, one geometric, the other algebraic. On the geometric side any element $\omega \in \omega_G^0(Y)$ can be represented as a proper fiber preserving

G map

1.2 $$\omega : Y \times \Omega \to Y \times \Omega$$

for a suitably large G module Ω. Moreover, any ω represented in this way is properly G homotopic to a map k transverse to $Y = Y \times 0 \subset Y \times \Omega$. It is important to specify the properties of k and the way in which k is properly homotopic to Y. This is done in [15] §8 and [13]. On the algebraic side, for the one point space q, $\omega_G^0(q) = \omega_G^0$ is the Burnside ring of G i.e. the Grothendieck group of finite G sets [14]. This important fact is taken up further later.

If $A \subset Y$ is an invariant subset and $\omega \in \omega_G^0(Y)$, the restriction of ω to A is denoted by ω_A. If $\omega \in \omega_G^0$ is viewed as a proper G map $\omega : \Omega \to \Omega$ and H is a subgroup of G written $H \subseteq G$, define an integer $\deg_H \omega$ by

1.3 $$\deg_H \omega = \text{degree } \omega^H$$

where $\omega^H : \Omega^H \to \Omega^H$ is the induced map on the H fixed point set of Ω. Then $\deg_H$ is a ring homomorphism into Z and the product of these over all $H \subseteq G$ defines an injective ring homomorphism to $\prod_{\mathcal{S}} Z$ where $\mathcal{S} = \mathcal{S}(G)$ is the set of subgroups of G. See [19].

Define a partial order on $\mathcal{S}(G)$ by $H \leq K$ <u>if</u> H <u>contains</u> K. Let G act on $\mathcal{S}(G)$ by conjugation of subgroups. Then $\mathcal{S}(G)$ is a G set and the action of G preserves the partial order i.e. $\mathcal{S}(G)$ is a G poset - a partially ordered G set.

In the following $\mathcal{K} \subset \mathcal{S}(G)$ is a G invariant subset containing 1. Define an ideal $J(G,\mathcal{K})$ and a multiplicatively closed subset $K(G,\mathcal{K})$ in ω_G^0 by

1.4 $$J(G,\mathcal{K}) = \ker\left(\omega_G^0 \xrightarrow{\text{Res}} \prod_{H \in \mathcal{K}} \omega_H^0\right)$$

1.5 $$K(G,\mathcal{K}) = J(G,\mathcal{K}) \cap \deg_G^{-1}(1)$$

Here $\mathrm{Res}_H : \omega_G^0 \to \omega_H^0$ denotes restriction to H . Let RO(G)(R(G)) denote the real (complex) representation ring of G . Define an ideal $I(G,\mathcal{X})$ in RO(G) by

1.6
$$I(G,\mathcal{X}) = \ker(RO(G) \xrightarrow{\mathrm{Res}} \prod_{H \in \mathcal{X}} RO(H))$$

We slightly abuse notation denoting by $\mathcal{P}$ all groups of prime power order and all subgroups of G of prime power order when we have a particular G in mind. The ideal $I(G,\mathcal{P})$ plays an especially significant role. The Adams operations may be used to construct elements in $I(G,\mathcal{P})$. In fact if p^n is the order of the p Sylow subgroup of G , then

1.7
$$\prod_{p \mid |G|} (\Psi^{p^n} - 1) \cdot RO(G) \subset I(G,\mathcal{P}) .$$

The geometrical problems posed in section 0 single out families of subgroups $\mathcal{X} \in \mathcal{S}(G)$ which in turn specify the algebraic objects $K(G,\mathcal{X})$ and $I(G,\mathcal{X})$ in terms of which the problems are solved.

§2 Answers to the semi-classical questions

The answers to the list of questions in 0 are expressed through a description of the structure of those groups which affirmatively answer each question. Let $\mathcal{G}^1$ denote the class of groups which extend a cyclic group by a p group i.e. $H \in \mathcal{G}^1$ if there is a normal p subgroup of H having cyclic quotient. If G is any group, $\mathcal{X}^1(G) = \mathcal{X}^1$ is the set of subgroups of G which are in $\mathcal{G}^1$. Let $\mathcal{G}$ be the class of groups G such that $K(G,\mathcal{X}^1)$ is not empty. For the remainder of this section G is <u>abelian</u>.

2.1
$$\mathcal{G}_1 = \{G \mid H \subset \mathcal{X}^1(G) \Rightarrow G/H \notin \mathcal{P}\}$$

2.2
$$\mathcal{G}_2 = \{G \mid G/P \in \mathcal{G} \text{ when } P \in \mathcal{P}\}$$

Note that $\mathcal{S}_2 \subset \mathcal{S}_1$.

Theorem 2.3. *Let* G *be an abelian group of odd order in* $\mathcal{S}_1$. *Then there is a smooth action of* G *on a closed homotopy sphere* Σ *with* Σ^G *one point* [15].

Theorem 2.4. *Let* G *be an abelian group of odd order in* $\mathcal{S}_2$. *Then there is a smooth action of* G *on a closed homotopy sphere* Σ *with* $\Sigma^G = p \cup q$ *and the isotropy representations* $T\Sigma_p$ *and* $T\Sigma_q$ *at* p *and* q *are distinct* [16].

More precisely suppose A and B are real G modules with $A^G = B^G = 0$, $A - B \in I(G,\mathcal{P})$, both representations satisfy the gap hypothesis 7.0 and $\dim A^P = \dim B^P \geq 5$ for $P \in \mathcal{P}$.

Theorem 2.5. *Let* G *be as in Theorem* 2.4. *Then there is a smooth action of* G *on a homotopy sphere* Σ *with* $\Sigma^G = p \cup q$ $T\Sigma_p = A$ *and* $T\Sigma_q = B$ *are the isotropy representations of* G *at* p *and* q [16].

Theorem 2.6. *Let* $G \notin \mathcal{P}$, *then there is a* G *module* A, *a smooth closed homotopy sphere* Σ *with* G *action, an* h *equivalence* $f : \Sigma \to S(A)$ *and* $\dim \Sigma^H = \dim S(A)^H$ $H \neq G$, $\Sigma^G = \phi$, $S(A)^G \neq \phi$. *If* G *is perfect*, Σ^H *is a homotopy sphere and* f^H *is a homotopy equivalence for* $H \neq G$ [16].

Theorem 2.7. *Let* $G \notin \mathcal{P}$ *and* V *be a complex* G *module with* $V^G = 0$. *Let* $\mathcal{H} = \mathrm{Iso}(V - 0)$ *and* $K(G,\mathcal{H})$ *be nonempty. Then there is a smooth action of* G *on a* $2n$ *dimension closed manifold* X *with* $X^G =$ *one point and isotropy representation* V [16].

Remark 2.8. If G is cyclic and not in $\mathcal{P}$, $K(G,\{1\})$ is not empty and there are complex G modules V with $\mathrm{Iso}(V - 0) = \{1\}$. In particular G acts smoothly on a closed 2 dimensional manifold with one fixed point.

§3 Realization Theorems

Although the statements of the theorems of section 2 are concerned with the components of the points fixed by G, their analysis requires the combinatorial structure of the set of components of fixed sets of all subgroups of G. This structure pervades the entire theory of G surgery and is essential even for such simple spaces as spheres and disks where surprisingly rich structure can occur for nonlinear actions. With this motivation we first describe the structure of the component set. Then we discuss two realization theorems which deal with realizing a given component structure of the fixed set of a G manifold and with realizing given normal representations at the fixed point sets.

If Y is a G space, let $\Pi(Y) = \coprod_{H \subseteq G} \pi_0(Y^H)$. Then $\Pi(Y)$ is a G poset. The partial order is defined like this: Let $|\alpha|$ denote the topological space defined by the component $\alpha \in \Pi(Y)$. If $\alpha, \beta \in \Pi(Y)$ define $\alpha \leq \beta$ if $|\alpha| \subset |\beta|$ and $K \leq H$ if $\alpha \in \pi_0(Y^K)$ and $\beta \in \pi_0(Y^H)$. In particular for $Y = q$, the one point space, $\Pi(q) = \mathcal{S}(G)$ with its partial order.

There is a G poset map $\rho : \Pi(Y) \to \mathcal{S}(G)$ defined by $\rho(\alpha) = H$ if $\alpha \in \pi_0(Y^H)$. This commutes with the G action and preserves partial order. For $\alpha \in \Pi(Y)$ let

3.1 $$G^\alpha = \bigcap_{x \in |\alpha|} G_x$$

where G_x is the isotropy group of the point x.

If $f : X \to Y$ is a G map, there is an induced map $\tilde{f}^H : \pi_0(X^H) \to \pi_0(Y^H)$ for $H \subseteq G$ and this defines a G map $\tilde{f} : \Pi(X) \to \Pi(Y)$. Let $\mathcal{N} \subset \mathcal{S}(G)$ and set

3.2 $$\Pi(X, \mathcal{N}) = \rho^{-1}(\mathcal{N}) .$$

It follows from Smith Theory that if f is an h equivalence, $\tilde{f} : \Pi(X, \mathcal{P}) \to \Pi(Y, \mathcal{P})$ is a homeomorphism. This fact motivates the frequently incorporated hypothesis that $\tilde{f}$ induces a homeomorphism. See 7.3.

Let Y be a smooth G manifold and $\alpha \in \Pi(Y)$. Set

3.3 $$G_\alpha = \{g \in G \mid g\alpha = \alpha\}.$$

The normal bundle $\nu(|\alpha|,Y)$ of $|\alpha|$ in Y is a G_α vector bundle over $|\alpha|$. For any $q \in |\alpha|$, the representation of G^α on the fiber of $\nu(|\alpha|,Y)$ at q is denoted by $s(\alpha)$ since it does not depend on q. Explicitly

3.4 $$s(\alpha) = \nu(|\alpha|,Y)_q \quad q \in |\alpha|.$$

There are three steps involved in the theorems of section 2. The first is to choose a model Y which serves an approximation to the statement of the theorem. E.g. in 2.3 choose a G module V, with $V^G = 0$. Then $Y = S(V \oplus \mathbb{R})$ has two fixed points and is a homotopy sphere. The second step is a geometric one. It produces a G map $f: X \to Y$ where X is a G manifold with the properties specified by the theorem except for the homotopy type. E.g. X^G = one point. The map f has properties to be specified precisely later. In particular the degree of f is one. The further conditions deal with the tangent bundle of X and the normal bundles $\nu(|\alpha|,X)$ for $\alpha \in \Pi(X)$. This data is relevant to the third step which is both algebraic and geometric. It is a process which converts (X,f) into (X',f') where f' is an h equivalence. This is the subject of the G surgery theory encompassed by 0.1.

Relevant to the second step we provisionally define a weak (framed) h normal map $\mathcal{V} = (X,f;Y)$ to be a G map $f: X \to Y$ between smooth manifolds of the same dimension such that degree f is one ($f^*TY = TX$ in $KO_G(X)$) and $\tilde{f}: \Pi(X,\mathcal{P}) \to \Pi(Y,\mathcal{P})$ is a homeomorphism. The precise definitions occur in §7.

Let $\Gamma \subset \pi_0(Y^G)$ and define a family $\mathcal{K}(\Gamma) \subset \mathcal{S}(G)$ by

3.5 $$\mathcal{K}(\Gamma) = \mathrm{Iso}(Y - Y^G) \qquad \text{if} \quad \Gamma \neq \pi_0(Y^G) \text{ or } \phi$$
$$= 1 \qquad \text{otherwise}$$

Define a multiplicative subset of ω_G^0 by

3.6 $$K(G,\Gamma) = K(G,\mathcal{K}(\Gamma))$$

This multiplicative subset derived from Y and the subset Γ of $\pi_0(Y^G)$ is the appropriate algebraic tool to treat the theorems of section 2 by means of the Realization Theorems I and II discussed now.

Suppose $\mathscr{U} = (X,f\,;Y)$ is a framed h normal map and $\Gamma \subset \pi_0(Y^G)$.

Definition: Γ is realized by $\mathscr{U}$ if $\tilde{f}$ maps $\pi_0(X^G)$ homeomorphically onto Γ and $\pi_0(X^H)$ surjectively onto $\pi_0(Y^H)$ for $H \neq G$. If there is some $\mathscr{U}$ with this property, we say Γ is realized.

Realization Theorem I 3.7. Suppose that $G^{\alpha} = P$ for all $\alpha \in \pi_0(Y^P)$ and $P \in \mathscr{P}$. If $\Gamma \subset \pi_0(Y^G)$ and $K(G,\Gamma) \neq \phi$, Γ can be realized.

Let $\mathscr{U} = (X,f\,;Y)$ be a weak h normal map which gives a homeomorphism $\tilde{f} : \pi_0(X^G) \to \pi_0(Y^G)$. Define $\gamma(\mathscr{U}) : \pi_0(Y^G) \to RO(G)$ by

$$\gamma(\mathscr{U})(\beta) = s(\beta) - s(\tilde{f}^{-1}(\beta)) \qquad \beta \in \pi_0(Y^G)$$

Definition: $z \in RO(G)$ is realized at $\beta \in \pi_0(Y^G)$ by $\mathscr{U}$ if $\gamma(\mathscr{U})(\alpha) = 0$ for $\alpha \neq \beta$ and $\gamma(\mathscr{U})(\beta) = z$. If there exists some $\mathscr{U}$ with this property, we say z is realized at β.

Realization Theorem II 3.8. Suppose for all $\alpha \in \pi_0(Y^P)$ that $G^{\alpha} = P$ whenever $P \in \mathscr{P}$. Let $\beta \in \pi_0(Y^G)$ and Γ be the complement of β. Suppose $K(G,\Gamma)$ is nonempty and $z = A - B \in I(G,\mathscr{P})$ where $A = s(\beta)$ and $\dim z^G = 0$, then z is realized at β.

Realization Theorem I is a consequence of: (i) a localization lemma relating $\omega_G^0(Y)$ and $\omega_G^0(Y^G)$ through $K(G,\Gamma)$ and (ii) a transversality theorem which produces framed h normal maps from certain elements in $\omega_G^0(Y)$. An outline of the proof of the Realization Theorem I is provided by the next four points 3.9-3.12.

If Σ is any subset of $\Pi(Y)$, let $|\Sigma|$ denote its underlying topological space.

Lemma 3.9. Let Γ be a proper nonempty subset of $\pi_0(Y^G)$ and Γ' its complement. If $e \in K(G,\Gamma)$ and $t = \{e^n, n = 0,1,2 \cdots\} \subset \omega_G^0$, the inclusion of $Y^G = |\Gamma \amalg \Gamma'|$ in Y induces an isomorphism of the localizations

$$t^{-1}\omega_G^0(Y) \to t^{-1}\omega_G^0(|\Gamma|) \times t^{-1}\omega_G^0(|\Gamma'|) = t^{-1}\omega_G^0(Y^G).$$

Proof: By definition $\mathrm{Res}_H(e) = 0$ for $H \in \mathrm{Iso}(Y - Y^G)$. Thus $e \cdot \omega_G^0(G/H) = e \cdot \omega_H^0 = 0$. Since Y is obtained from Y^G by adding cells of type G/H for $H \in \mathrm{Iso}(Y - Y^G)$, the result follows.

Corollary 3.10. Let Γ be a subset of $\pi_0(Y^G)$ with complement Γ'. If $K(G,\Gamma)$ is nonempty, there is an $\omega \in \omega_G^0(Y)$ with $\omega_{|\Gamma|} = 1$ and $\deg_G \omega_q = 0$, $\deg_1 \omega_q = 1$ for all $q \in |\Gamma'|$.

Proof: Let $e \in K(G,\Gamma)$ and $\omega = 1 - \theta$, where θ is in the ideal $e \cdot \omega_G^0(Y)$ and θ restricts to $(0, e^\lambda) \in \omega_G^0(|\Gamma|) \times \omega_G^0(|\Gamma'|)$ for some $\lambda > 0$. The existence of θ follows from 3.9 when Γ is a proper nonempty subset. If $\Gamma = \pi_0(Y^G)$ take $\mathscr{V} = 1$ and if $\Gamma = \phi$ take $\omega = 1 - e$.

Theorem 3.11. Let Y be a smooth G manifold and $\Gamma \amalg \Gamma'$ a decomposition of $\pi_0(Y^G)$. Let $\omega \in \omega_G^0(Y)$ where $\omega_{|\Gamma|} = 1$ and $\deg_G(\omega_q) = 0$ $\deg_1(\omega_q) = 1$ for all $q \in |\Gamma'|$. Then ω is properly G homotopic to a map k transverse to Y with $X = k^{-1}(Y)$ and if $f = k|X : X \to Y$, there is a framed h normal map $\mathscr{V} = (X, f; Y)$ where $f^G : X^G \to |\Gamma|$ is a diffeomorphism. See [13] and 1.2.

Remark 3.12. If $G^\alpha = P$ for all $P \in \mathscr{P}$, we can suppose that $\tilde{f}$ defines a homeomorphism from $\pi_0(X^P)$ to $\pi_0(Y^P)$ for all $P \in \mathscr{P}$ and a surjective map of $\pi_0(X^H)$ onto $\pi_0(Y^H)$ for $H \neq G$.

Realization Theorem I is a corollary of 3.9-3.12. In fact the Γ of 3.9 is realized by the $\mathscr{V}$ of 3.11.

Realization Theorem II is actually a consequence of the Realization Theorem I. Let β be a component of Y^G and Γ the union of the remaining components of Y^G. Let $A = s(\beta)$ and B be a G module with $B^G = 0$ and $A - B \in I(G,\mathscr{P})$. The dimension of $|\beta|$ is d. Suppose $K(G,\Gamma) \neq \emptyset$. Let X be the manifold obtained by realizing Γ using 3.7 and $f : X \to Y$ the resulting map.

If $d > 0$ let $X_\beta = S(B \oplus R^{d+1})$. If $d = 0$, let $Y_\beta = S(B \oplus R)$; so $Y_\beta^G = p \cup q$. Let $\Gamma_1 = p \in \pi_0(Y_\beta^G)$. Apply Realization Theorem I to Y_β and realize Γ_1 by producing a G manifold X_β with X_β^G consisting of one point p' with $s(p') = B$. This is possible because if $K(G,\Gamma) \neq \emptyset$, then $K(G,\Gamma_1) \neq \emptyset$. In either case we have a G manifold X_β with $\pi_0(X_\beta^G)$ consisting of one point α with $s(\alpha) = B$.

If $Z = X \amalg X_\beta$ is the disjoint union and $F : Z \to Y$ is defined by $F_{|X} = f$ and $f_{|X_\beta}$ maps X_β to one point in $|\beta|$, then there is a <u>weak</u> h normal map $\mathcal{H} = (Z,F;Y)$ where $\tilde{F} : \pi_0(Z^G) \to \pi_0(Y^G)$ is a homeomorphism, $\gamma(\mathcal{H})(\beta) = A - B$ and $\gamma(\mathcal{H})(\alpha) = 0$ for $\alpha \neq \beta$. This is clear from construction and the fact that $f^* TY = TX$. At this point $\tilde{F}$ does not define a homeomorphism from $\Pi(Z,\mathcal{P})$ to $\Pi(Y,\mathcal{P})$. This is rectified by zero dimensional G surgery. (See [15] §4-5.) The last point requires that $A - B \in I(G,\mathcal{P})$, f induces a homeomorphism from $\Pi(X,\mathcal{P})$ to $\Pi(Y,\mathcal{P})$ and there is such a homeomorphism from $\Pi(X_\beta,\mathcal{P})$ to $\Pi(S(B \oplus R^{d+1}),\mathcal{P})$. This completes the outline of Realization Theorem II.

§4 Realization Theorems and the Theorems of §2

The first step in proving 2.3-2.7 is to select a model which approximates the statement of the theorem. The model for these theorems is of the form $Y = S(V \oplus R^n)$ where V is a complex G module with $V^G = 0$ and $\mathrm{Iso}(V - 0) = \mathcal{K}$ specified so $K(G,\mathcal{K}) \neq \emptyset$. The integer n is one except for 2.6 and there n is even and not zero.

The second step is to produce a weak (framed) h normal map $\mathcal{H} = (X,f;Y)$ where X satisfies all the requirements of the relevant theorem in §2 except the condition that X be a homotopy sphere. This step requires the Realization Theorems and which is explained in this section. The last step is to convert $(X,f;Y)$ to $(X',f';Y)$ where f' is an h equivalence. It is treated by applying the sequence 0.1. This is discussed in later sections.

The discussion of the application of the second step to theorems 2.3-2.7 follows.

(i) Theorem 2.3. Let $G \in \mathscr{S}_1$ and V be a complex G module with $V^G = 0$ and $\mathrm{Iso}(V - 0) = \{H \subseteq G \mid G/H \notin \mathscr{P}\}$. If $Y = S(V \oplus \mathbb{R})$, then $Y^G = p \cup q$ consists of two isolated fixed points. Let $\Gamma = p$ and $\Gamma' = q$ be the decomposition of $\pi_0(Y^G) = Y^G$. Then $K(G,\Gamma)$ is nonempty because $G \in \mathscr{S}_1$. (Note that $\mathrm{Iso}(V - 0) = \mathrm{Iso}(Y - Y^G)$.) Apply 3.7 to obtain a framed h normal map $\mathscr{U} = (X, f\,; Y)$ where $\tilde{f}$ maps $\pi_0(X^G)$ homeomorphically on Γ. Since $TX = f^*TY$ in $KO_G(X)$ and X^G is connected, X^G consists of one point p' with $f(p') = p$ and $TX_{p'} = V$ is the isotropy representation of G at p'.

(ii) Theorems 2.4 and 2.5. Let $G \in \mathscr{S}_2 \subset \mathscr{S}_1$ and A and B be two distinct G modules which satisfy the hypothesis for V in (i) such that $A - B \in I(G,\mathscr{P})$. Let $Y = S(A \oplus R)$ so $Y^G = p \cup q$. Choosing the same decomposition of $\pi_0(Y^G)$ as above, $K(G,\Gamma)$ is nonempty so 3.8 applies. There is a weak h normal map $\mathscr{U} = (X, f\,; Y)$ such that $\tilde{f}$ induces a homeomorphism of $\pi_0(X^G)$ on $\pi_0(Y^G)$ and if $X^G = p' \cup q'$, then $TX_{p'} = TY_p = A$ while $TX_{q'} - TY_q = B - A$. In particular the isotropy representations at p' and q' are distinct.

(iii) Theorem 2.6. Let $G \notin \mathscr{P}$ and V be a complex G module with $V^G = 0$, and $\mathrm{Iso}(V) = \mathscr{S}(G)$. Let n be even and $Y = S(V \oplus R^n)$. Then $Y^G = S(\mathbb{R}^n)$ is connected (and dimension Y^H is odd for all $H \in G$. See §10 for relevance of this). Let $\Gamma = \emptyset$. Then $K(G,\Gamma)$ is the subset of ω_G^0 consisting of elements e with $\deg_G(e) = 0$ and $\deg_1(e) = 1$. Since $G \notin \mathscr{P}$, $K(G,\Gamma)$ is nonempty. By 3.7, there is a framed h normal map $\mathscr{U} = (X, f\,; Y)$ where $\pi_0(X^G) = \emptyset$ while $\tilde{f}$ maps $\pi_0(X^H)$ onto $\pi_0(Y^H)$ for $H \neq G$. Since $\mathscr{U}$ is a framed h normal map, $\dim X^H = \dim Y^H$ for $H \neq G$ also $X^G = \emptyset \neq Y^G$ so $\dim X^G \neq \dim Y^G$.

(iv) Theorem 2.7. Suppose $G \notin \mathscr{P}$. Let V be a complex G module with the properties prescribed in Theorem 2.7. Let $Y = S(V \oplus \mathbb{R})$ so $\mathrm{Iso}(Y - Y^G) = \mathrm{Iso}(V - 0)$. Let $\Gamma = p$, $\Gamma' = q$ where $Y^G = p \cup q$. The hypotheses of Theorem 2.7 imply that $K(G,\Gamma) \neq \emptyset$. Apply Theorem 3.7 to produce a framed h normal map $\mathscr{H} = (X,f\,;Y)$ where X^G consists of one point p' with isotropy representation V.

§5 Combinatorics of Smooth G Manifolds

This section develops the combinatorial aspects of a smooth G manifold pertinent to the terms in the exact sequence 0.1. Specifically a G Poset pair is defined. The parameter λ in the exact sequence is a G Poset pair. The group structure and functorial structure of $I^h(G,\lambda)$, indeed its definition, depends on the properties of λ.

For any G space Y, the G poset $\Pi(Y)$ has been defined. When Y is a smooth G manifold, this set has additional structure: In particular it contains an invariant G poset $\pi(Y)$ which is a more efficient record of the G invariant of Y. In addition to the G poset map $\rho : \Pi(Y) \to \mathcal{S}(G)$, there are these additional invariants of the G action on Y: Let $\alpha \in \Pi(Y)$

5.1 $$d : \Pi(Y) \to Z^+ \quad \text{(non-negative integers)}$$
$$d(\alpha) = \dim |\alpha| . \quad \text{To emphasize } Y \text{ set } d = d_Y .$$

5.2 $$G^\alpha = \bigcap_{y \in |\alpha|} G_y \quad \text{- the principle isotropy group of } |\alpha| .$$

5.3 $s(\alpha)$ - the representation of G^α on any normal fiber of $|\alpha|$ in Y. To emphasize Y, set $s = s_Y$.

We say Y is <u>oriented</u> if for each $\alpha \in \Pi(Y)$, we have chosen a generator $[|\alpha|] \in H_{d(\alpha)}(|\alpha|,\partial|\alpha|,Z) \cong Z$. Define $-Y$ to be the oriented manifold defined by Y but with associated generator $-[|\alpha|]$ for $\alpha \in \Pi(Y)$. In the case Y is oriented and $\alpha \in \Pi(Y)$ with $\rho(\alpha) = G^\alpha$, there is a multiplicative homomorphism

5.4 $$w(\alpha) : W(\alpha) \to \{-1,1\} \subset Z \quad \text{where}$$
$$W(\alpha) = G_\alpha/G^\alpha$$

This is defined by $w(\alpha)(g)[|\alpha|] = g_*[|\alpha|]$ for $g \in W(\alpha)$. Note that $|\alpha|$ is a $W(\alpha)$ manifold.

There is a redundance in the data of $\Pi(Y)$ which is more efficiently carried by

5.5 $$\pi(Y) = \{\alpha \in \Pi(Y) \mid \rho(\alpha) = G^{\alpha}\}$$

An important property of the G poset which derives from the fact that Y is a <u>smooth G manifold</u> is this: For $\alpha \in \pi(Y)$ let $\pi_{\alpha} = \{\beta \in \pi(Y) \mid \beta \geq \alpha\}$. Then if $K \geq \rho(\alpha)$ there is a unique maximal element in the set $\{\beta \in \pi_{\alpha} \mid K \geq \rho(\beta)\}$. This follows from the fact that $\rho(\pi_{\alpha}) = \mathrm{Iso}(s(\alpha))$ and $\mathrm{Iso}(s(\alpha))$ is closed under intersections. There is a retraction $r : \Pi(Y) \to \pi(Y)$ defined by $r(\alpha) =$ unique maximal element in $\{\beta \in \pi(Y) \mid \beta \leq \alpha\}$. In fact $r(\alpha) = \beta$ where $|\beta| = |\alpha|$ and $\rho(\beta) = G^{\alpha}$. Note r is a G poset map and preserves the function d.

Now suppose X and Y are smooth oriented G manifolds and $f : X \to Y$ is a G map. Define $\hat{f} : \pi(X) \to \pi(Y)$ by the composition $\pi(X) \to \Pi(X) \xrightarrow{\tilde{f}} \Pi(Y) \xrightarrow{r} \pi(Y)$. Then $\hat{f}$ is a G poset map.

When Σ is a subset of $\Pi(X)$, denote by $|\Sigma|$ its underlying topological space and by $f_{\Sigma} : |\Sigma| \to |\tilde{f}(\Sigma)|$ the restriction of f to $|\Sigma|$. When $\alpha \in \Pi(Y)$ and $\Sigma = \tilde{f}^{-1}(\alpha)$, f_{Σ} is written f_{α}. <u>In particular</u> f_{α} <u>is defined for</u> α <u>in</u> $\Pi(X)$ <u>or</u> $\Pi(Y)$. On the set of $\alpha \in \Pi(Y)$ such that the dimension of each element of $\tilde{f}^{-1}(\alpha)$ equals that of α, define an integral valued function $\deg \hat{f}$ by

5.6 $$\deg \hat{f}(\alpha) = \text{degree } f_{\alpha}$$

and set

5.7 $$\pi(f) = (\pi(X), \pi(Y), \hat{f}, \deg \hat{f})$$

<u>Example 1</u>. Let $G = Z_{pq}$ be the cyclic group of order pq where p and q are relatively prime integers. Choose positive integers a and b so that $-ap + bq = 1$. Let N respectively M be the complex 2 dimensional Z_{pq}

modules defined by

$$N : t(z_0,z_1) = (t^p \cdot z_0, t^q \cdot z_1)$$

$$M : t(z_0,z_1) = (t \cdot z_0, t^{pq} z_1)$$

where $(z_0,z_1) \in N$ respectively M are complex coordinates and $t = \exp 2\pi i/pq$ generates Z_{pq}. Define a G map $k : N \to M$ by

$$k(z_0,z_1) = (\bar{z}_0^a z_1^b, z_0^q + z_1^p)$$

This induces a G map $f : S(N) \to S(M)$ by $f(z) = k(z) \cdot \|k(z)\|^{-1}$ where $z = (z_0,z_1)$ and $\|w\|$ denotes the norm of w. Computation shows that degree $f = 1$ so f is an h equivalence.

These points should be emphasized: $\mathrm{Iso}(S(N)) = \{1, Z_p, Z_q\}$ and $\mathrm{Iso}(S(M)) = \{1, Z_{pq}\}$ are distinct. In particular $\mathrm{Iso}(\cdot)$ is not an invariant of an h equivalence. The degrees of the maps f^H on H fixed sets are (q,p) for $H = Z_p$ respectively Z_q and q is a unit mod p while p is a unit mod q. The G normal bundle of $S(N)^H$ in $S(N)$ is the pull back via f^H of some G vector bundle over $S(M)^H$ which is not its normal bundle in $S(M)$. This last fact motivates the data of an ambient G map in §7.

Abbreviate $S(N)$ and $S(M)$ by X and Y so $f : X \to Y$. In these cases ρ is injective on $\pi(X)$ and $\pi(Y)$ with image $\mathrm{Iso}(X)$ respectively $\mathrm{Iso}(Y)$. View ρ as an identification. Then these G posets together with $\hat{f}$ are described by:

$$\mathrm{Iso}(X) \quad \begin{array}{ccc} & 1 & \\ \rotatebox{45}{$\leq$} & & \rotatebox{-45}{$\geq$} \\ Z_p & & Z_q \end{array} \qquad \begin{array}{c} 1 \\ \rotatebox{90}{$\leq$} \\ Z_{pq} \end{array}$$

$$\hat{f}(1) = 1, \ \hat{f}(H) = Z_{pq} \quad \text{for } H = Z_p \text{ or } Z_q,$$

$$\deg \hat{f}(1) = 1, \ \deg \hat{f}(Z_p) = q, \ \deg \hat{f}(Z_q) = p.$$

We want to abstract the properties of the data $\pi(f)$. Let π be a G poset possessing a G poset map ρ to $\mathcal{S}(G)$. Let $\Sigma \subset \pi$ and set

5.8 $$\pi_\Sigma = \left\{\beta \in \pi \mid \beta \geq \alpha \text{ for some } \alpha \in \Sigma\right\}$$

5.9 $$\mathcal{S}(G)_\Sigma = \left\{K \in \mathcal{S}(G) \mid K \geq \rho(\alpha) \text{ some } \alpha \in \Sigma\right\}$$

5.10 $$G_\Sigma = \left\{g \in G \mid g\,\Sigma = \Sigma\right\}$$

5.11 $$W(\Sigma) = G_\Sigma/\rho(\Sigma) \text{ - defined whenever } \rho \text{ is constant on } \Sigma .$$

Remark. If ρ is constant on Σ, then $\rho(\Sigma)$ is a normal subgroup of G_Σ so $W(\Sigma)$ is a group. This is mainly used when $\Sigma = \sigma$ consists of one point of π.

Henceforth all G posets are equipped with a unique maximal element m and a G poset map $\rho : \pi \to \mathcal{S}(G)$ with $\rho(m) = 1$ - the maximal element of $\mathcal{S}(G)$; moreover, we require for each $\alpha \in \pi$ that $\rho : \pi_\alpha \to \mathcal{S}(G)_\alpha$ is an injection of posets and if $\alpha \leq \beta$ and $\beta \leq \alpha$, then $\alpha = \beta$. If $\rho\pi_\alpha = \mathcal{S}(G)_\alpha$ for all α, we say π is complete. A G poset π with the property that for each $\alpha \in \pi$ there is a unique maximal element in the set $\{\beta \in \pi_\alpha \mid K \geq \rho(\beta)\}$ whenever $K \in \mathcal{S}(G)_\alpha$ is called smooth. We say $\pi \subset \Pi$ is dense if for each $\alpha \in \Pi$ there is a unique maximal element $r(\alpha)$ in the set $\{\beta \in \pi \mid \beta \leq \alpha\}$. By way of motivation $\Pi(Y)$ is complete and $\pi(Y) \subset \Pi(Y)$ is a dense subset whenever Y is a smooth (connected) G manifold. Any smooth G poset π has a unique completion $\bar{\pi}$ which is a functor of π in the sense that if $\tau : \pi \to \pi'$ is a G poset map between smooth G posets, there is a unique G poset map $\bar{\tau} : \bar{\pi} \to \bar{\pi}'$ with $\rho = \rho'\bar{\tau}$. In particular $\Pi(Y) = \overline{\pi(Y)}$ and $\tilde{f} = \bar{\hat{f}}$. See [15] §1.

5.12 Definition: A G Poset π consists of a G poset π together with functions d, s and w where

(i) $d : \pi \to Z^+$ is a G map preserving partial order

(ii) $s(\alpha)$ is a real representation of $\rho(\alpha)$ for $\alpha \in \pi$

(iii) $w(\alpha) : W(\alpha) \to \{-1,1\} \subseteq Z$ is a multiplicative homomorphism for $\alpha \in \pi$.

Note that a smooth G manifold Y defines a G Poset $\pi(Y)$ using 5.1-5.4.

A G Poset map is by definition a map of the underlying G posets. A G Poset Map is a G poset map τ which preserves s, w and ρ and $G_{\tau(\alpha)} = G_\alpha$. A G Poset Map whose underlying mapping is an isomorphism, surjection or injection of sets is called a G Poset isomorphism, surjection or injection. If d is also preserved by a G Poset isomorphism, it is called a G Poset equivalence. A G Poset isomorphism is denoted by $\cong$.

Example 2. The most primitive but still important example of a G Poset π is given by $\pi = m$. This G Poset has one element m with $\rho(m) = 1$. The associated data consists of an integer $d = d(m)$, $s(m) = 0$ and a homomorphism $w(m) : G \to Z_2$.

5.13 Definition: A G Poset pair λ written $\lambda = (\pi_1, \pi_2, \tau, \mu)$ consists of smooth G Posets π_i $i = 0, 1$ together with a G Poset map $\tau : \pi_1 \to \pi_2$ and a function μ from $\bar{\pi}_2^*$ to the non-negative integers Z^+. Here

$$\bar{\pi}_2^* = \left\{\alpha \in \bar{\pi}_2 \,|\, d_1 r(\beta) = d_2 r(\alpha) \text{ for } \beta \in \bar{\tau}^{-1}(\alpha)\right\}$$

Example 3. Let X and Y be smooth oriented G manifolds and $f : X \to Y$ be a G map. Set $\pi(f) = (\pi(X), \pi(Y); \hat{f}, \deg \hat{f})$. Then $\pi(f) = \lambda$ is a G poset pair if $\deg \hat{f}(\alpha) \geq 0$ whenever it is defined. This is the canonical example of a G Poset pair. It is exactly this data which gives rise to the groups $I^c(G,\lambda)$.

Let $\lambda^i = (\pi_1^i, \pi_2^i, \tau^i, \mu^i)$ be two G Poset pairs $i = 1, 2$. A map of G Poset pairs $\iota : \lambda^1 \to \lambda^2$ is a pair of G Poset maps $\iota_j : \pi_j^1 \to \pi_j^2$ for $j = 1, 2$ such that $\tau^2 \iota_1 = \iota_2 \tau^1$. It is a G Poset pair isomorphism surjection or injection if and only if both ι_1 and ι_2 are G Poset isomorphisms, surjections or injections. The chief example of a map of G poset pairs arises like this: Let $W_0 \subset W$, $Z_0 \subset Z$ be G invariant submanifolds of W and Z. Let $F : (W,W_0) \to (Z,Z_0)$ with $F_0 = F|W_0$. The inclusions $\iota_1 : W_0 \to W$ and $\iota_2 : Z_0 \to Z$ give rise to a map of G Poset pairs $\iota : \pi(F_0) \to \pi(F)$ where $\iota = (\hat{\iota}_1, \hat{\iota}_2)$.

<u>Remark</u>. Suppose W is a smooth G manifold and $W_0 \subset \partial W$ is a smooth G invariant submanifold. The inclusion ι of W_0 in W induces a G Poset map $\hat{\iota} : \pi(W_0) \to \pi(W)$. This map commutes with ρ and preserves s by the smooth G collar neighborhood Theorem [4]. Without some further hypothesis, it need not happen that $G_{\hat{\iota}(\alpha)} = G_\alpha$ for $\alpha \in \pi(W_0)$. The hypothesis that $\hat{\iota}$ be a G Poset injection surjection or equivalence guarantees this.

§6 G Surgery and $\tilde{K}_0(Z(G))$

In order to simplify the exposition, we make the following assumption for all G manifolds:

6.1 $$\rho\pi(X,\mathcal{P}) = \mathcal{P} \qquad X \text{ a smooth } G \text{ manifold.}$$

If $\alpha \in \pi(X,P)$ and P is a p group different from 1, set $R_\alpha = Z_{(p)}$ the integers localized at p. If $P = 1$, so $\alpha = m$, set $R_m = Z$.

Let $f : A \to B$ be a G map and set $K_n(f,R) = \mathrm{Ker}(H_n(A,R) \xrightarrow{f_*} H_n(B,R))$. It is an $R(G) = \Lambda$ module. If for each n it is a projective Λ module, set

6.2 $$\chi(f,R) = \sum (-1)^{n+1} K_n(f,R) \in \tilde{K}_0(\Lambda)$$

A specific occurrence of this function appears when $\mathcal{W} = (X,f;Y)$ is a (weak) h normal map as follows: Suppose $\alpha \in \pi(X,\mathcal{P})$ and

6.3 (i) $K_*(f_\beta,R_\beta) = 0$ for $\beta < \alpha$ $\beta \in \pi(X,\mathcal{P})$

(ii) $K_i(f_\alpha,R_\alpha) = 0$ unless $i = n$.

Then $K_n(f_\alpha, R_\alpha)$ is a projective $R_\alpha(G) = \Lambda_\alpha$ module and we set

6.4 $$\chi_\alpha(f) = \chi(f_\alpha, R_\alpha) \in \tilde{K}_0(\Lambda_\alpha) .$$

The assumptions of 6.3 are natural from two points of view. If f is an h equivalence, they are satisfied as a consequence of Smith Theory. In addition they form the basis for the inductive process (on the partial order in $\pi(X,\mathcal{P})$) for converting f to an h equivalence.

In order to motivate the subsequent treatment of these "obstructions" consider the invariant $\chi(f,Q)$. Since $Q(G)$ is semisimple, every $Q(G)$ module is projective so $\chi(f,Q)$ is always defined. In fact if $f: A \to B$ and M_f denotes the mapping cone,

$$\chi(f,Q) = \sum (-1)^n \tilde{H}_n(M_f, Q) .$$

Let $\Omega(G)$ denote the set of equivalence classes of compact G cell complexes under the equivalence relation defined by $X \sim Y$ if $\chi(X^H) = \chi(Y^H)$ for all $H \subseteq G$. Here $\chi(\cdot)$ denotes Euler characteristic. See [19]. A zero dimensional G manifold is a finite G set so there is a natural homomorphism (with respect to disjoint union) of the Grothendieck group of finite G sets into $\Omega(G)$ which is an isomorphism. There is a homomorphism from $\Omega(G)$ to $\tilde{K}_0(Q(G))$ defined for $X \in \Omega(G)$ by

$$\gamma_{G,Q}(X) = \sum (-1)^i H_i(X,Q) .$$

It is evident from the above discussion that $\chi(f,Q) = \gamma_{G,Q}(M_f) - 1 = \gamma_{G,Q}(B - A)$; so if $\Delta(G,Q)$ is the kernel of $\gamma_{G,Q}$, then $\chi(f,Q)$ vanishes if and only if $A - B \in \Delta(G,Q)$.

In a similar but more complicated manner, the obstruction $\chi_\alpha(f)$ encountered for an h normal map and dealing with the ring Z and its localization $Z_{(p)}$ can be treated in the framework of $\Omega(G)$. In particular there is an ideal $\Delta(G) = \Delta(G,Z) \subset \Omega(G)$ such that the obstructions $\chi_\alpha(f)$ (if defined) vanish for $\alpha \in \pi(X,\mathcal{P})$ $\alpha \neq m$ and $\chi_m(f) \in B_0(G) \subset \tilde{K}_0(Z(G))$ if and only if $Y - X \in \Delta(G)$. See [12], [13]. Here $B_0(G)$ is a geometrically defined subgroup of the reduced projective class group.

The geometric process of G surgery deals with the isotropy groups of the source manifold X where $\mathcal{H} = (X,f;Y)$ is an h normal map. At the G homotopy level it is related to the operation $X \mapsto X \cup G/H \times D^i$ where $H \subset Iso(X)$ and $G/H \times S^{i-1} \subset X$ is an invariant subset. This operation changes X in $\Omega(G)$ to $X + (-1)^i G/H$. Let $\mathcal{K} \subset \mathcal{S}(G)$ be a G invariant subset and

6.5 $$\Omega(G,\mathcal{K}) = \text{subgroup of } \Omega(G) \text{ generated by } G/H \text{ for } H \in \mathcal{K}.$$

Then for $\mathcal{K} = Iso(X)$, the operation in $\Omega(G)$ $X \mapsto X \cup G/H \times D^i$ is the identity modulo $\Omega(G,\mathcal{K})$.

This discussion indicates that the subgroup $\Delta(G)$ is not a G surgery invariant but that $\Delta(G) + \Omega(G,\mathcal{K})$ is for a suitably defined family $\mathcal{K} \subset \mathcal{S}(G)$. The choice of $\mathcal{K}$ depends on the geometry.

Let $\lambda = (\pi_1,\pi_2,\tau,\mu)$ be a G Poset pair. Set

$$\mathcal{K}_\lambda = \{H \subseteq G \mid H = \rho(\alpha), \alpha \in \pi_1, d(\alpha) > 0\}$$

$$\Psi(\lambda) = \Delta(G) + \Omega(G,\mathcal{K}_\lambda).$$

Let $\mathcal{H} = (X,f;Y)$ and $\pi(f) = \lambda$. Set

6.6 $$\theta(f) = Y - X \in \Omega(G)/\Psi(\lambda) = \text{\textcircled{H}}(\lambda)$$

<u>Lemma 6.7</u>. <u>$\theta(f)$ is a G surgery invariant and vanishes if f is an h equivalence</u>.

§7 The Terms in the Exact Sequence 0.1

Throughout the remainder of the text, we impose the following condition on all G manifolds:

7.0 Gap hypothesis for X: For each $\alpha \in \pi(X,\mathcal{P})$ require $\max\{d(\beta) \mid \beta < \alpha, \beta \in \pi(X)\} < \frac{1}{2} d(\alpha)$.

Let $\lambda = (\pi_1, \pi_2, \tau, \mu)$ be a G Poset pair where $\tau : \pi_1 \to \pi_2$ and $\mu : \vec{\pi}_2^* \to Z^+$. Require λ to satisfy:

7.1 $\vec{\pi}_2^* \supset \bar{\pi}_2(\mathcal{P})$. See 5.13. Here $\bar{\pi}(\mathcal{P}) = \rho^{-1}(\mathcal{P})$, $\rho : \bar{\pi} : \to \mathcal{S}(G)$.

7.2 $\mu(\bar{\tau}(\alpha))$ is a unit of R_α for $\alpha \in \bar{\pi}_1(\mathcal{P})$.

7.3 $\bar{\tau} : \bar{\pi}_1(\mathcal{P}) \to \pi_2(\mathcal{P})$ is a homeomorphism.

These requirements are imposed if $\lambda = \pi(f)$ and f is an h equivalence. Henceforth all G Poset pairs are required to satisfy 7.1-7.3 unless otherwise stated.

A G triad $(X;X_0,X_1)$ consists of a G manifold X with $\partial X = X_0 \cup X_1$ and $X_0 \cap X_1 = \partial X_0 = \partial X_1$. The submanifolds X_0 and X_1 of ∂X are G invariant. A G map of triads preserves this structure. If $\Sigma \in \pi(X)$, then Σ_0 denotes the set of components of $X_0 \cap |\Sigma|$; so $\Sigma_0 \subset \pi(X_0)$.

Let $F_i : W_i \to Z$ be two G maps for $i = 0, 1$. If there is a G manifold pair (Q,P) and a G map $F : (Q,P) \to (Z \times I, \partial Z \times I)$ such that $\partial Q = W_0 \cup W_1 \cup P$, $\partial W_0 \cup \partial W_1 = \partial P = P \cap (W_0 \cup W_1)$, $F|W_i = F_i$ where $F|W_i = W_i \to Z \times i$, we say $(Q;P,F)$ is a G cobordism between (W_0,F_0) and (W_1,F_1). If $P = X \times I$, $X \times 0 = \partial W_0$, $X \times 1 = \partial W_1$ and F preserves I coordinate we say the cobordism is relative boundary. We shall be concerned with various additional structures on such G maps. These induce corresponding notions of cobordism where the additional structure is required to extend over the cobordism.

If X is a G space, ξ is a G vector bundle over X and $K \subset G$ acts trivially on X, then ξ splits as an $N(K)$ (normalizer of K) vector bundle $\xi = \xi^K \oplus \xi_K$ where ξ^K is the K fixed set and ξ_K is the orthogonal complement of ξ^K. If η is a G vector bundle over a G space Y, $f : X \to Y$ is a G map and $b : \xi \to f^*\eta$ is a G vector bundle isomorphism, it splits as $b = b^K \oplus b_K$ where $b^K : \xi^K \to (f^K)^* \eta^K$ and $b_K : (f^K)^* \xi_K$ are $N(K)$ vector bundle isomorphisms.

Let π be a G poset contained in $\Pi(X)$ where X is a G space. A <u>(weak)</u> π <u>vector bundle</u> ξ over X is by definition a collection $\{\xi_\alpha | \alpha \in \pi\}$ where each ξ_α is a $G_\alpha(\rho(\alpha))$ vector bundle over $|\alpha|$ such that if $\rho(\beta) = K$,

7.4 $\quad \xi_\beta = (\xi_\beta)_K$ and if $\alpha \leq \beta$ then

$\xi_{\beta}\big|_{|\alpha|} = (\xi_\alpha)_K$. Note $G_\alpha \cap G_\beta = G_\alpha \cap N(K)$.

7.5 $\quad$ For each $g \in G$, there are vector bundle isomorphisms $\hat{g} : \xi_\alpha \to \xi_{g\alpha}$ such that $\hat{g}_1 \hat{g}_2 = \widehat{g_1 g_2}$ and if $g \in G_\alpha(\rho(\alpha))$, $\hat{g}$ is the isomorphism given with the structure of ξ_α as a $G_\alpha(\rho(\alpha))$ vector bundle.

A (weak) π <u>vector bundle map</u> $b : \xi \to \eta$ is a collection of $G_\alpha(\rho(\alpha))$ vector bundle maps $b_\alpha : \xi_\alpha \to \eta_\alpha$ $\alpha \in \pi$ such that

7.6 $\quad$ If $\alpha \leq \beta$ $(b_\alpha)_K = (b_\beta)\big|_{|\alpha|}$, when $\rho(\beta) = K$

7.7 $\quad b_{g\alpha} = \hat{g} b_\alpha \hat{g}^{-1}$ for $g \in G$.

<u>Remark</u>. If $b : \xi \to \eta$ is a (weak) π vector bundle map, ker b and coker b are defined in an obvious way and are (weak) π vector bundles.

<u>Examples</u>.

1) If N is a G vector bundle over a G space X define $N_\alpha = (N_{||\alpha|})_{\rho(\alpha)}$ for $\alpha \in \Pi(X)$. This gives a $\Pi(X)$ vector bundle over X denoted by N.

2) If X is a smooth G manifold, define $\nu(\cdot,X)$ to be the $\Pi(X)$ vector bundle defined by $\nu(\cdot,X)_\alpha = \nu(|\alpha|,X)$ - the G_α normal bundle of $|\alpha|$ in X.

If $f:X \to Y$ is a G map and ξ is a $\Pi(Y)$ vector bundle over Y, define a $\Pi(X)$ vector bundle $f^!(\xi)$ by

$$f^!(\xi)_\alpha = f_\alpha^*(\xi_\beta) \quad \text{where} \quad \tilde{f}(\alpha) = \beta .$$

If ξ is a $\Pi(X)$ vector bundle and $\pi \subset \Pi(X)$, it defines a π vector bundle by restriction. By a slight abuse of notation this is also called ξ. A fact implicit in the following is that for a smooth G manifold X, there is a natural equivalence between $\pi(X)$ and $\Pi(X)$ vector bundles.

An <u>ambient</u> G <u>map</u> $\mathcal{H} = (W,F,b,C)$ consists of a G map $F:W \to Z$ between smooth G manifolds together with a stable G vector bundle ξ over Z and a $\Pi(Z)$ bundle η over Z, a stable G vector bundle isomorphism $C: TW \to F^*\xi$ and a $\Pi(W)$ vector bundle isomorphism $b: \nu(\cdot,W) \to F^!\eta$. The stable $\pi(W)$ vector bundle isomorphism determined by b is denoted by B.

A <u>framed</u> <u>ambient</u> G <u>map</u> $\mathcal{H}$ requires ξ to be TY. A <u>weak</u> <u>ambient</u> G <u>map</u> $\mathcal{H}$ requires only that C be a stable vector bundle isomorphism (not necessarily compatible with G) and that b be a weak $\Pi(X,\mathcal{P})$ vector bundle map.

A (framed-weak) ambient G map of triads $\mathcal{H} = (W,F,b,C)$ is a (framed-weak) ambient G map such that $F:W \to Z$ is a G map of triads. Denote by $\mathcal{H}_0$ the ambient G map (W_0,F_0,b_0,C_0) where F_0, b_0 and C_0 are the restrictions of F, b, C to W_0. Similarly define $\mathcal{H}_1$.

A (<u>weak</u>-<u>framed</u>) <u>h</u> <u>normal</u> <u>map</u> $\mathcal{H} = (W,F,b,C)$ is a (weak-framed) ambient G map between oriented G manifolds such that for each $\alpha \in \pi(W,\mathcal{P})$ and $P \in \mathcal{P}$ (i) $d(\alpha) = d(\hat{F}(\alpha))$ (ii) $\hat{F}$ defines a surjection of $\pi(W,P)$ onto $\pi(Z,P)$ (iii) For each $\beta \in \Pi(Z,\mathcal{P})$ deg $\hat{F}(\beta)$ is a unit of R_β.

<u>Remarks</u>. This is the definiton of an h normal map which appears in [15]. In earlier sections this was abbreviated by $\mathcal{H} = (W,F;Z)$ and we continue this where useful. Also condition (ii) was strengthened to require $\hat{F}$ to define an

isomorphism of $\pi(W,P)$ onto $\pi(Z,P)$. The condition that $\rho\pi(Z,\mathcal{P}) = \mathcal{P}$ guarantees that this can be arranged if $\hat{F}$ is surjective onto $\pi(Z,P)$ for $P \in \mathcal{P}$. With this strengthened condition $\pi(F) = \lambda$ satisfies 7.1-7.3.

If $F : W \to Z$ is a G map between smooth G manifolds, we assume it maps boundary to boundary and let $\partial F : \partial W \to \partial Z$ be the induced map on boundaries. A <u>(weak-framed) h map</u> $\mathcal{W} = (W,F;Z)$ is a (weak-framed) h normal map such that $\hat{F}$ and $\partial\hat{F}$ induces homeomorphisms on $\pi(\cdot,P)$ for $P \in \mathcal{P}$ and $\theta(F) = Y - X = 0$ in $\mathcal{H}(\pi(F))$ (6.6). A (weak-framed) h Equivalence is a (weak-framed) h map as above with F an h equivalence. For notational convience we omit reference to the (weak-framed) terminology.

An <u>h normal cobordism</u> between two h normal maps $\mathcal{W}_i = (X_i,f_i,b_i,C_i)$ $f_i : W_i \to Z$ is an h normal map $\mathcal{W} = (W,F,\tilde{b},\tilde{C})$ with $\partial W = X_1 \cup X_2 \cup P$, $\tilde{b}_i|X_i = b_i$ $\tilde{C}|_{X_i} = C_i$ and (W,P,F) is a G cobordism between (X_i,f_i) $i = 1, 2$.

An <u>h cobordism</u> between h maps $\mathcal{W}_i = (X_i,f_i,b_i,C_i)$ $i = 1, 2$ is an h normal cobordism $\mathcal{W} = (W,F,\tilde{b},\tilde{C})$ between them where $\mathcal{W}$ is an h map and the inclusions $X_i \to W$ induce G Poset isomorphisms $\pi(X_i) \to \pi(W)$ for $i = 1, 2$.

An <u>h Equivalence cobordism</u> between two h Equivalences is an h cobordism between them which itself is an h Equivalence.

The following material generalizes the procedure of [20] Chapter 9 for defining surgery groups.

Let $\mathcal{I}^h(G,\lambda)$ denote the set of h maps $x = (X,f,b,C)$ with $\pi(f) = \lambda$ and ∂f an h equivalence. Denote the h map $(-X,f,b,C)$ by $-x$. Specifically the orientations of both source and target are reversed in $-x$. Define an equivalence relation $\sim$ on $\mathcal{I}^h(G,\lambda)$ as follows: Let $x_i = (X_i,f_i,b_i,C_i)$ $i = 1, 2$. Then $x_1 \sim x_2$ if there is an h map of triads $\mathcal{W} = (W,F,\tilde{b},\tilde{C})$ with (i) $W_0 = X_1 - X_2$, $F_0 = f_1 \amalg f_2 : X_1 - X_2 \to Y_1 - Y_2$, $\tilde{b}_0 = b_1 \amalg b_2$, $\tilde{C}_0 = C_1 \amalg C_2$, (ii) F_1 is an h equivalence (iii) The inclusions of X_i in W and Y_i in Z induce an isomorphism $\pi(f_i) \to \pi(F) = \lambda$. Then

7.11 $$I^h(G,\lambda) = \mathcal{I}^h(G,\lambda)/\sim$$

This set contains a distinguished element 0. It is represented by any h map $(X,f,b,C) = \mathcal{W}'$ which occurs as $\mathcal{W}_0$ for some h map of triads $\mathcal{W} = (W,F,\widetilde{b},\widetilde{C})$ such that F_1 is an h Equivalence, $\lambda = \pi(F_0) \to \pi(F)$ is a G Poset pair isomorphism.

Let $\eta_G^h(Y,\lambda)$ denote the subset of $\mathcal{J}^h(G,\lambda)$ consisting of those h maps whose target is Y. An equivalence relation $\sim$ on $\eta_G^h(Y,\lambda)$ is defined by saying $x_i = (X_i,f_i,b_i,C_i)$ for $i = 1,2$ are equivalent if x_1 and x_2 an h cobordant relative boundary. Set

7.12 $$N_G^h(Y,\lambda) = \eta_G^h(Y,\lambda)/\sim$$

Let $h\mathcal{S}_G^h(Y,\lambda)$ denote the subset of $\eta_G^h(Y,\lambda)$ consisting of h Equivalences. An equivalence relation is defined by h Equivalence cobordism relative boundary. Set

7.13 $$hS_G^h(Y,\lambda) = h\mathcal{S}_G^h(Y,\lambda)/\sim$$

The inclusions

$$h\mathcal{S}_G^h(Y,\lambda) \subset \eta_G^h(Y,\lambda) \subset \mathcal{J}^h(G,\lambda)$$

give rise to the sequence

7.14 $$hS_G^h(Y,\lambda) \xrightarrow{d} N_G^h(Y,\lambda) \xrightarrow{\sigma} I^h(G,\lambda)$$

and clearly $\sigma^{-1}(0) \supset \text{image } d$. In fact the sequence is exact i.e. $\sigma^{-1}(0) = \text{image } d$.

If $\lambda = (\pi_1,\pi_2,\tau,\mu)$ where $\pi_i = \underline{m}$, $\tau(m) = m$, $\mu(m) = 1$, $d_i(m) = d$ for $i = 1,2$ and $\omega_i(m)$ are homomorphisms of $W(m) = G$ to Z_2, then $I^h(G,\lambda)$ is the Wall group $L_d(Z(G))$ of the group ring $Z(G)$ with involution defined by $\omega(m) = \omega_1(m)\omega_2(m)$. See §5, Example 2.

The group structure on $I^h(G,\lambda)$ is rather subtle. In fact it is not always a group.

Theorem 7.15 [7]. Let $\lambda = \pi_1, \pi_2, \tau, \mu)$ be a G Poset pair. Let d_i be the dimension functions (5.1) for π_i $i = 0, 1$. If $d_1(\alpha) \geqq 6$ $\alpha \in \pi_1$ and if $d_1(\alpha) \leqq d_2(\tau(\alpha))$ for all α, then $I^c(G,\lambda)$ is an abelian group.

To emphasize the content of this theorem, we note it is easy to construct an h equivalence $f : X \to Y$ and an h equivalence $g : Y \to X$ such that, if $\lambda = \pi(f)$, $I^h(G,\lambda)$ is a group but $I^h(G,\lambda')$ is not where $\lambda' = \pi(g)$.

Whether or not $I^h(G,\lambda)$ is a group, there is a well defined restriction function $\mathrm{Res}_H : I^h(G,\lambda) \to I^h(H,\lambda|_H)$ for $H \subset G$. It is a homomorphism when these are both groups.

The parameter h can be replaced by fh or wh denoting the framed h maps or weak h maps. There is a natural map between the sequences of 7.14 for fh, h and wh acknowledging a weaking of hypotheses. Let c denote one of these conditions.

§8 Constructing Elements in $N_G^h(Y,\lambda)$

One useful and general method for producing elements in $N_G^h(Y,\lambda)$ is provided from the relation between G transversality, $\omega_G^0(Y)$ and $\Omega(G)$ ([15]§8). Very briefly here is a discussion:

Let $\omega \in \omega_G^0(Y)$ and suppose $\deg_1 \omega_q = 1$ for all $q \in Y$. Then a transversality result like 3.11 (see [15] §8 and [13]) produces a smooth G manifold X and a degree one map $f : X \to Y$. If $\rho\pi(Y,\mathcal{P}) = \mathcal{P}$, we may suppose $\hat{f}$ induces a homeomorphism on $\pi(\cdot,P)$ for $P \in \mathcal{P}$. So $\mathcal{U} = (X,f;Y)$ is an h normal map. Let $\pi(f) = \lambda$.

In order that $\mathcal{U}$ be an h map i.e. $\mathcal{U} \in N_G^h(Y,\lambda)$, we need $\theta(f) = 0$ in $\vdash\!\!\theta\!\!\dashv(\lambda)$. This gives a condition on ω as $\mathcal{U}$ is a function of ω. It remains a problem to give a general description of $\theta(f)$ in terms of ω. Under specific conditions this is done in [13] §9. Two important points in this regard are (i) $\omega_G^0(Y^G) = \omega_G^0 \otimes_Z \omega_1^0(Y^G)$ where 1 denotes the trivial group and (ii) ω_G^0 is isomorphic to $\Omega(G)$. Since $\vdash\!\!\theta\!\!\dashv(\lambda)$ lies in a quotient group of $\Omega(G)$, it is natural to try to express $\theta(f)$ as a function of the restriction $\omega|_{Y^G}$ of ω to Y^G.

(This lies in $\Omega(G) \otimes_Z \omega_1^0(Y^G)$.) When Y^G consists of isolated points this is done in [13] §9.

Actually the h maps produced from $\omega_G^0(Y)$ lie in $N_G^{fh}(Y,\lambda)$ for various λ depending on the choice of $\omega \in \omega_G^0(Y)$. The Realization Theorem 3.7 and its consequences 2.3, 2.6 and 2.7 depend on this construction of elements in $N_G^{fh}(Y,\lambda)$. The Realization Theorem 3.8 and its consequences 2.4 and 2.5 depend on the weaker notion of a weak h normal map and the construction of elements in $N_G^{wh}(Y,\lambda)$. One method of construction is outlined in the outline of the proof of the Realization Theorem 3.8. See also [16].

§9 Functorial Properties of $I^c(G,\lambda)$

There are two important properties of the groups (sets) $I^c(G,\lambda)$ involved in geometric application in particular to the application of the theorems of section 2. They are a finiteness theorem and an induction theorem.

Let $\lambda = (\pi_1,\pi_2,\tau,\mu)$ be a G Poset pair. For $\alpha \in \pi_1(\mathcal{P}) = r\rho^{-1}(\mathcal{P})$ $(\rho : \bar{\pi}_1 \to \mathcal{S}(G))$ and $\mathcal{Y} = (X,f;Y) \in I^c(G,\lambda)$, let

$$\mathrm{Sign}_\alpha(\mathcal{Y}) = \mathrm{Sign}(W(\alpha),|\alpha_x|) - \mathrm{Sign}(W(\alpha),|\hat{f}(\alpha_x)|)$$

where $\alpha_x \in \pi(X,\mathcal{P})$ corresponds to α under the isomorphism of $\pi(f)$ with λ and $\mathrm{Sign}(G,Z)$ is the Atiyah-Singer G signature of the G manifold Z. It is a character of G i.e. an element of $R(G)$. This construction defines a homomorphism (when $I^c(G,\lambda)$ is a group)

9.1 $$\mathrm{Sign} : I^c(G,\lambda) \longrightarrow \prod_{\alpha\in\pi_1(\mathcal{P})} R(W(\alpha)).$$ Set

9.2 $$I_0^c(G,\lambda) = \ker(\mathrm{Sign})$$

<u>Theorem 9.3</u> [7]. $I_0^c(G,\lambda) \otimes Q = 0$

This is the finiteness theorem. It holds for c = wh, fh or h and indeed can be generalized to a wider class of conditions.

Corollary 9.4. Let $\lambda = (\pi_1, \pi_2, \tau, \mu)$ be a G Poset pair, $d(\alpha) \geq 6$ and $d(\alpha)$ odd whenever $\alpha \in \pi_1$ or π_2. Then $I^c(G,\lambda) \otimes Q = 0$.

If $F(\cdot)$ is a contravariant functor of groups, let $\mathrm{Res}_H : F(G) \to F(H)$ denote the restriction homomorphism induced by the inclusion of H in G. In particular we have $\mathrm{Res}_H : I^c(G,\lambda) \longrightarrow I^c(G,\lambda_{|H})$. This is a group homomorphism whenever $I^c(G,\lambda)$ is a group. Here $\lambda_{|H}$ denotes the restriction to H of the G Poset pair λ to an H Poset pair.

One very useful type of zero theorem for the groups (sets) $I^c(G,\lambda)$ is an induction theorem which describes the minimal collection $\mathcal{K}_\lambda$ of subgroups of G such that

9.5 $$0 \to I^c(G,\lambda) \xrightarrow{\mathrm{Res}} \coprod_{H \in \mathcal{K}_\lambda} I^c(H,\lambda_{|H})$$

is exact. Here $(\mathrm{Res})_H = \mathrm{Res}_H$. Of course such a theorem is of no use if $\mathcal{K}_\lambda$ contains G. For groups of odd order this is done [7]. Here is a simply stated version.

Theorem 9.6 [7]. Let G be a group of odd order and $\lambda = (\pi_1, \pi_2, \tau, \mu)$ be a G poset pair with $\rho\pi_2(\mathcal{P}) = \mathcal{P}$. Let $\mathcal{K}_\lambda$ denote the set of subgroups of G which are either hyperelementary or are in $\mathcal{S}^1$. Then the sequence 9.5 is an exact sequence of groups (sets).

Let A be a real G module with $A^G \neq 0$. Let $Y = S(A)$. Suppose λ is a G poset pair with $\lambda = (\pi_1, \pi_2, \tau, \mu)$, $\pi_2 = \pi(Y)$ and $d(\alpha) \geq 6$ for $\alpha \in \pi_1$ or π_2.

Theorem 9.7 [6]. $N^c_G(Y,\lambda)$ is an abelian monoid and $\sigma : N^c_G(Y,\lambda) \to I^c(G,\lambda)$ is a homomorphism.

Remark. This is a surprising fact. The hypothesis implies $Y^G \neq \emptyset$ so an addition is defined by connected sum in the target of an h map $\mathcal{Y} = (X,f;Y)$ with $\pi(f) = \lambda$. However, X^G may well be empty depending on λ and so connected sum in the source is not defined.

§10 Completing the Proofs of §2

In §4 we began the outline of the proofs of the theorems of §2. Starting from $Y = S(V\oplus R^n)$ we constructed a (weak-framed) h normal map $\mathcal{N} = (X,f;Y)$ where X has all the properties required by the relevant theorem except its homotopy type. In one manner or another $\mathcal{N}$ was produced from the equivariant cohomology theory $\omega_G^0(\cdot)$ and the multiplicative set $K(G,\Gamma)$. In particular the construction was arranged so that $\theta(f) = 0$; so $\mathcal{N} \in N_G^c(Y,\lambda)$ for $c = wh$ or fh and λ a suitable G Poset pair. Except for the case of 2.6, this is an intricate point and discussion must be restricted to the comment that the families $\mathcal{S}_1$ and $\mathcal{S}_2$ in 2.3-2.5 were chosen so that the obstruction $\theta(\cdot)$ would vanish. In the case of 2.6, the framed h normal map $\mathcal{N} = (X,f;Y)$ has the property that X^H and Y^H are closed odd dimensional manifolds for all $H \subseteq G$; so both X and Y represent 0 in $\Omega(G)$ as $\chi(X^H) = \chi(Y^H) = 0$ for all H. This implies $\theta(f)$ is zero by 6.6.

Consider first the case of 2.6. Section 4 provides an $\mathcal{N} = (X,f;Y) \in N_G^h(Y,\lambda)$ for suitable λ. Since λ carries all the information required for $\dim X^H$ and $\dim Y^H$ and is chosen so that $\dim Y^H - \dim X^H = 0$ for $H \neq G$ and is non-zero for $H = G$, it suffices to find an $\omega' \in hS_G^h(Y,\lambda)$. For this case n is even and greater than or equal to 8 and $\dim |\alpha|$ is odd for any $\alpha \in \pi(X)$ or $\pi(Y)$. By 9.3 $I_0^h(G,\lambda) = I^h(G,\lambda)$ is a torsion group. It follows from 9.7 that some multiple $k\mathcal{N}$ of $\mathcal{N}$ is in the kernal of σ in the exact sequence of 7.14. This means $k\mathcal{N} = d(\mathcal{N}')$ for some $\mathcal{N}' \in hS^h(Y,\lambda)$. Then $\mathcal{N}' = (\Sigma,f';Y)$ and Σ satisfies the claim of 2.6.

The case 2.3: Starting from $\omega = 1 - \theta \in \omega_G^0(Y)$ with $Y = S(V\oplus R)$ and θ in the ideal $e \cdot \omega_G^0(Y)$ $e \in K(G,\Gamma)$, we construct an $\mathcal{N} = (X,f;Y) \in N_G^h(Y,\lambda)$ for suitable λ. (See §4 i and the proof of 3.10.) Because θ is in the ideal $e \cdot \omega_G^0(Y)$, $\mathrm{Res}_H(\mathcal{N}) = 1$ for all H which are hyperelementary. This means that $\mathrm{Res}_H(\mathcal{N}) = i_H^*(1)$ where $1 \in \omega_H^0(Z)$, Z is the unit disk of $V \oplus R$ and $i_H : Y \to Z$ is the inclusion. Using this fact construct an H map $\mathcal{N}(H) = (W(H),f(H);Z)$ with $\mathcal{N}_0(H) = (X,f;Y) = (\partial W(H),\partial f(H),\partial Z)$ using an H transversality arguement as in [15] §8 and [13]. Viewing $W(H)$ and Z as H triads with e.g. $W(H)_0 = \partial W(H)$ and $W(H)_1 = \phi$, we see that $\mathrm{Res}_H\, \sigma(\mathcal{N})$ is zero in $I^h(H,\lambda_{|H})$ by definition. (It must be shown that $\pi(f)_{|H} \to \pi(F(H))$ is a G Poset pair isomorphism.) Since this holds for all hyperelementary groups and since G is abelian, 9.6 applies and $\sigma(\mathcal{N})$ is zero; so $\mathcal{N} = d(\mathcal{N}')$ for some $\mathcal{N}' = (\Sigma, f', Y) \in hS^h(Y,\lambda)$ with Σ^G = one point as dictated by λ.

The cases 2.4 and 2.5. The principle is the same as in the preceding discussion. §4ii produces an $\mathcal{N} = (X,f;Y) \in N_G^{wh}(Y,\lambda)$ with $Y = S(V \oplus R)$ and λ suitable. The induction Theorem 9.6 is again applied to conclude that $\mathcal{N} = d(\mathcal{N}')$ for some $\mathcal{N}' \in hS^h(Y,\lambda)$.

§11 Motivating the Definition of Zero in $I^c(G,\lambda)$

The importance of the zero element in $I^c(G,\lambda)$ is clear from the definitions of the terms in the exact sequence 7.14. Here is a restatement of the essential assumptions for $\mathcal{N}' = (X,f;Y) \in I^c(G,\lambda)$ to be zero. There must be a c normal map of triads $\mathcal{N} = (W,F;Z)$ with $\mathcal{N}_0 = \mathcal{N}'$ and

11.1 F_1 is a c equivalence

11.2 $\pi(F_0) \to \pi(F)$ is a G Poset pair isomorphism

11.3 $\theta(F) = 0$

11.4 All relevant manifolds satisfy the Gap hypothesis.

The condition 11.1 needs no justification. The motivation of 11.3 has been done in §6. We discuss 11.2 and 11.4.

The relevance of zero in $I^c(G,\lambda)$ is due to the fact that if $\mathcal{N} \in N_G^c(Y,\lambda)$ and $\sigma(\mathcal{N}) = 0$, then $\mathcal{N} = d(\mathcal{N}')$ for some $\mathcal{N}' \in hS_G^c(Y,\lambda)$. Without the inclusion of 11.2 in the definition of zero this exactness property of 7.14 is false. Here is an example:

Let $G = Z_2$ be the cyclic group of order 2 with generator g and V the complex $2n$ dimensional G module defined by $gv = -v$ for $v \in V$. Then G

acts freely on $Y = S(V)$ and with one fixed point 0 in $Z = D(V)$. The inclusion $Y \subset Z$, therefore, does not induce a G Poset Isomorphism of $\pi(Y)$ in $\pi(Z)$.

Let

$$W = \left\{(z_1, z_2 \cdots z_{2n+1}) \mid \sum_{i=1}^{2n} z_i^2 + z_{2n+1}^q = \varepsilon, \sum |z_i|^2 \leq 1\right\}$$

Here $\varepsilon > 0$ is small and real. It is a variety in C^{2n+1} which is invariant under the action of G defined by $g(v, z_{2n+1}) = (-v, z_{2n+1})$ for $v \in C^{2n}$ and $z_{2n+1} \in C$. Let $X = \partial W$. Then G acts freely on X and if q is odd, there is an h map $f : X \to Y$ which extends to an h normal map $\mathscr{W} = (W, F; Z)$.

Let $\lambda = \pi(f)$. Since G acts freely on X and Y, $I^c(G,\lambda) = L_d(Z(G))$ where $d = \dim X = 4n - 1$. Then $\mathscr{W}' = (X, f; Y) \in N_G^c(Y,\lambda)$ and $\sigma(\mathscr{W}') \in L_{4n-1}(Z(G))$. A computation due to Browder and Giffen shows this to be nonzero if q is $\pm 3 \bmod 8$. By the surgery product formula [20] $\sigma(\mathscr{W}' \times CP^{2k}) \in L_{4(n+k)-1}(Z(G))$ is nonzero where $\mathscr{W}' \times CP^{2k} = (X \times CP^{2k}, f \times 1d, Y \times CP^{2k})$ and CP^{2k} is complex projective space of real dimension $4k$. Then

$$\mathscr{W} \times CP^{2k} = (W \times CP^{2k}, F \times 1d, Z \times CP^{2k})$$

is an h normal map of triads if $W_0 = X$, $W_1 = \phi$, $Z_0 = Y$ and $Z_1 = \phi$. If n and k are properly chosen the gap hypothesis for all manifolds is satisfied. If $k > 0$, $\theta(F \times 1d) = 0$. Since $W_1 = Z_1 = \phi$, $F_1 \times 1d$ is a c equivalence. The essential point is that $\pi(F_0 \times 1d) \to \pi(F \times 1d)$ is not a G Poset pair isomorphism. Since $\sigma(\omega' \times CP^{2k}) \neq 0$ in $I^c(G,\lambda)$, this shows the necessity of 11.3.

The reason for the gap hypothesis is geometrically clear. The process of G surgery on X is involved with G imbeddings of the form $G/H \times S^i \to X$. This is the same as an imbedding of S^i in X^H which misses X^K whenever K contains H and $gS^i \cap S^i = \phi$ for g different from 1 lying in the normalizer of H. The gap hypothesis guarantees that the imbeddings relevant to G surgery exist. It can be removed by additionally supposing in the hypothesis of an h

map $\mathcal{N} = (X,f;Y)$ that $G_{f(x)} = G_x$ for $x \in X$ and that f is transverse to all fixed sets. See [5]. These assumptions are almost never satisfied for an h equivalence f and so render the theory with this hypothesis quite useless.

The paper [6] of Dovermann contains additional remarks about the gap hypothesis and its relevance to the algebra of G surgery. See also Remark 12.3.

§12 The Obstruction Theory

The entire theory encompassed by the exact sequence 7.14 is based on an obstruction theory outlined in [14] and [13]. Here is a brief description:

Let $\mathcal{N} = (X;f,Y)$ be an h normal map. A G invariant subset $\Sigma \subset \pi(X,\mathcal{P})$ is said to be closed if $\alpha \in \Sigma$, $\beta \in \pi(X,\mathcal{P})$ and $\beta \leq \alpha$ implies $\beta \in \Sigma$. Each $\alpha \in \pi(X,\mathcal{P})$ contributes two obstructions to converting $\mathcal{N}$ into an h equivalence.

12.1 $$\chi_\alpha(f) \in \tilde{K}_0(R_\alpha(W(\alpha))$$

12.2 $$\sigma_\alpha(f) \in L_{d(\alpha)}(R_\alpha(W(\alpha)))$$

These obstructions are defined and applied inductively with respect to the partial order in $\pi(X,\mathcal{P})$. Suppose $\Sigma \subset \pi(X,\mathcal{P})$ is closed, $\beta \in \pi(X,\mathcal{P})-\Sigma$ is a minimal element and f_α is an R_α homology equivalence for $\alpha \in \Sigma$. Then $\chi_\beta(f)$ is defined. If $\chi_\beta(f)$ vanishes, $\sigma_\beta(f)$ is defined. If $\sigma_\beta(f)$ vanishes, we may suppose f_β is an R_β homology equivalence; so f_α is a homology equivalence for $\alpha \in \Sigma \cup G\beta$.

<u>Remark 12.3</u>. The definition of $\sigma_\alpha(f)$ as an element of the Wall group $L_{d(\alpha)}(R_\alpha(W(\alpha)))$ requires the gap hypothesis. It cannot be defined otherwise [6].

The obstruction theory is crucially used in the proof that the sequence 7.14 is exact and in Theorems 9.3 and 9.6. The definition of zero in the group $I^c(G,\lambda)$ (§7) guarantees that the obstructions $\sigma_\alpha(f)$ and $\chi_\alpha(f)$ all vanish if $\mathcal{N} = (X,f;Y)$ is zero in this group. This is the main step in proving 7.14 is exact.

Theorems 9.3 and 9.6 are consequences of the existence of this obstruction theory and these properties of the target groups $\tilde{K}(\cdot)$ and $L(\cdot)$ of the obstruction $\chi_{\bullet}(f)$ and $\sigma_{\bullet}(f)$:

12.4 $\operatorname{Ker}(L_d(A(G)) \xrightarrow{\text{Sign}} R(G))$ is a torsion group. Here A is a subring of Q and $\text{Sign}\,\mathcal{Y} = \text{Sign}(G,Y)-\text{Sign}(G,X)$ if $\mathcal{Y} = (X,f;Y)$ is a geometric representation of an element of $L_d(A(G))$ [20].

12.5 $0 \to F(A(G)) \xrightarrow{\text{Res}} \prod_{H \in \mathcal{H}} F(A(H))$ is exact if G has odd order. Here F is either $\tilde{K}_0(\cdot)$ or $L_d(\cdot)$, A is Z or $Z_{(p)}$ p odd and $\mathcal{H}$ is the set of hyperelementary groups if A is Z and is the set of cyclic groups if A is $Z_{(p)}$ [8].

12.6 The Wall group $L_d(A(G))$ is zero whenever d is odd, $|G|$ is odd and A is Z or $Z_{(p)}$ p an odd prime [20].

BIBLIOGRAPHY

1. Atiyah, M.F. and Bott, R., The Lefschetz fixed point theorem for Elliptic Complexes II, Ann. of Math. 86(1967), 451-491.

2. ______, Notes on the Lefschetz fixed point theorem for Elliptic Complexes II, Notes, Harvard University (1964).

3. Borel,A., et. al., Seminar on transformation groups, Ann. of Math. Studies 46, Princeton University Press, (1960).

4. Bredon, G., Introduction to compact transformation groups, Academic Press, (1972).

5. Browder, W. and Quinn, F., A surgery theory for G manifolds and stratified sets, Manifolds-Tokyo, University of Tokyo Press, (1973).

6. Dovermann, H., Addition in G surgery groups, to appear.

7. Dovermann, H. and Petrie, T., G surgery III, to appear.

8. Dress, A., Induction and structure theorems for orthogonal representations of finite groups, Ann. of Math. 102(1975), 291-325.

9. Feit, W., Characters of finite groups, Benjamin, New York (1967).

10. Montgomery, D. and Samelson, H., Fiberings with singularities, Duke J. Math. 13(1946), 51-56.

11. Montgomery, D. and Yang, C.T., A generalization of Milnor's theorem and differentiable dihedral transformation groups, to appear.

12. Oliver, R. and Petrie, T., G surgery in the homotopy category and $K_0(Z(G))$, to appear. Proceedings of Northwestern Topology Conference, 1977.

13. Petrie, T., Pseudoequivalences of G manifolds, Proc. Sym. Pure Math., 32 (1977), 119-163.

14. ______, G maps and the projective class group, Comm. Math. Helv. 39(1977), 611-626.

15. ______, G Surgery II - Groups which act on a homotopy sphere with one fixed point, to appear.

16. ______, G Surgery IV - Semi-classical questions in transformation groups, to appear.

17. ______, G Surgery V, Infinite groups, to appear.

18. Petrie, T. and tom Dieck, T., Geometric modules over the Burnside ring, to appear.

19. tom Dieck, T., The Burnside ring of a compact Lie group I, Math. Ann. 215(1975), 235-250.

20. Wall, C.T.C., Surgery on compact manifolds, Academic Press, (1970).

SMOOTH CE MAPS AND SMOOTH HOMEOMORPHISMS

Martin Scharlemann

University of California, Santa Barbara

A CE map $f: X \to Y$ is a proper map such that each $f^{-1}(y)$, y in Y, has the shape of a point. Such maps have the property that, if X and Y are ANR's, then f is a homotopy equivalence. In fact, Siebenmann has shown that if X and Y are closed manifolds of dimension $n \geq 5$ and $\epsilon > 0$, then f is ϵ-homotopic, through CE maps, to a homeomorphism [Si]. Here we ask whether any smooth CE map $f: M \to N$ of smooth closed manifolds of dimension $n \geq 5$ is smoothly ϵ-homotopic through CE maps to a smooth homeomorphism. An obstruction arises in $H^3(N;\theta_3^h)$, where θ_3^h denotes the group of all homotopy 3-spheres modulo those bounding smooth contractible 4-manifolds. The theory is thus analogous to that of Cohen-Sullivan [CS] in the PL category. This is perhaps unexpected, for a PL CE map $f: X \to Y$ has the property that $f^{-1}(y)$ is a subcomplex of X, hence an ANR, hence contractible. On the other hand, smooth CE maps may have much wilder point inverses. For example, it is easy to construct a smooth proper map $f: R^2 \to R^2$ whose only non-degenerate point inverse is the Polish arc

$$\left\{ (x,y) \mid x = 0 \text{ and } |y| \leq 1 \text{ or } 0 < x \leq 1,\ y = \sin\frac{1}{x} \right\},$$

on which the derivative also vanishes.

Smooth homeomorphisms have been studied in [SS]. They differ from diffeomorphisms in allowing points on which the derivative vanishes. Up to isotopy, smooth homeomorphisms of smooth manifolds correspond to PL homeomorphisms between smooth triangulations of the manifolds.

In this paper I have tried to sacrifice both generality and precision for clarity. More detail and generalization will appear elsewhere [Sc_1] [Sc_2]. In particular, questions about rounding corners, connecting smooth maps together, and dealing with manifold boundaries are there treated in full.

§1. Smooth CE maps — some examples:

Let B^k denote the closed k-dimensional unit ball around $0 \in R^k$. For any $0 \le r < 1$, let $\mu_r: R \to [0,1]$ be a smooth map such that $\mu_r^{-1}(0) = (-\infty, r]$, $\mu_r(s) = 1$ for s near 1, and $\mu_r | \mu_r^{-1}(0,1)$ is a diffeomorphism, then $f_r: B^k \to B^k$ defined by $f_r(x) = \mu_r(|x|)x$ is a smooth CE map whose only non-degenerate point inverse is $f_r^{-1}(0) = rB^k$. In particular, f_0 is a smooth homeomorphism.

Actually, it is fairly easy to construct a smooth 1-parameter family of CE maps running from the identity to f_r for any r. In particular, if $g_0, g_1 M \to B^k$ are any two smooth CE maps which are equal over a neighborhood of ∂B^k, then, by composing with the above 1-parameter family of maps, we can construct a smooth 1-parameter family of CE maps from each g_i to a map which collapses the entire ball in B^k over which g_0 and g_1 may differ to a point. Hence we have

Lemma 1.1: Suppose $g_0: M \to B^k$, $g_1: M \to B^k$ are smooth CE maps with $g_0 = g_1$ over a neighborhood of ∂B^k. Then there is a smooth 1-parameter family of CE maps. $g_t: M \to B^k$ from g_0 to g_1.

We also have

Lemma 1.2: Any smooth homeomorphism $g : \partial B^k \to \partial B^k$ extends to a smooth homeomorphism $B^k \to B^k$.

Proof: The homeomorphism cone(g) is smooth except at 0. Compose cone(g) with f_0 to get a smooth homeomorphism with trivial derivative at 0.

§ 2. Finding a transverse triangulation

If $f: M \to N$ is any smooth map of closed smooth manifolds, any Whitehead triangulation [Wh] of N can be isotoped slightly so that the simplices of N are transverse to f, that is, so that each k-simplex may be extended to a smooth (non-proper) imbedding of R^k which is transverse to f. This is accomplished inductively as follows: Since regular values are dense, any 0-simplex is near a regular value, to which it is isotoped. Suppose f is transverse over the k - 1 skeleton and σ is a k-cell. Then the composition of f with the projection of a neighborhood of σ onto a copy of R^{n-k} orthogonal to σ has a regular value ρ near 0, the image of σ in R^{n-k}. The inverse image of ρ is then a k-simplex σ' (parallel to σ) over which f is transverse. The isotopy of σ to σ' covering a ray from 0 to ρ may be "tapered" near $\partial\sigma$ so that $\partial\sigma$ remains fixed during the isotopy. This completes the inductive step.

Suppose now that $f: M \to N$ is a smooth CE map between n-manifolds, n arbitrary, and suppose the 3 and 4 dimensional smooth Poincare conjectures were true. Choose any Whitehead triangulation of N and isotope it so that it is transverse to f, as described above. Then for any k-simplex σ in N, $f^{-1}(\sigma)$ is a k-manifold, hence an ANR. Since f is CE, $f^{-1}(\sigma)$ is a homotopy cell, hence (neglecting problems with corners), precisely B^k. If $k = 0$ then $f|B^0$ is a homeomorphism. Suppose inductively that f is a homeomorphism over a neighborhood of the $(k-1)$-skeleton N_{k-1} of N. Applying lemmas 1.1 and 1.2 to each k-cell (ignoring corners) there is an extension of $f|f^{-1}(N_{k-1})$ over the k-skeleton N_k, an extension which differs from $f|N_k$ by a smooth 1-parameter family of CE maps fixed near N_{k-1}. By tapering off the difference between this extension and $f|N_k$ into a neighborhood of N_k, we may alter f to a smooth CE map which is a homeomorphism over a neighborhood of $f^{-1}(N_k)$. By induction we have

Theorem 2.1. *Assume every smooth homotopy k-sphere, $k = 3,4$, is diffeomorphic to S^k. Then any smooth CE map $f: M \to N$ of smooth closed n-manifolds is smoothly homotopic through CE maps to a smooth homeomorphism.*

Remark: The homotopy may be taken as small as desired by choosing a sufficiently fine triangulation of N.

§ 3. The obstruction co-cycle

In the previous section we saw that any smooth CE map $f: M \to N$ between closed manifolds may be smoothly homotoped through CE maps to a map which is a homeomorphism over a 2-skeleton N_2 of N. Furthermore, the inverse image of any 3-simplex σ will be a homotopy 3-cell. Since $f|f^{-1}(\partial\sigma) \to \partial\sigma$ is a homeomorphism, there is a natural way to cap off the 3-cell to get a homotopy 3-sphere.

Thus to any 3-simplex in N is associated a homotopy 3-sphere, and thus an element of θ_3^h. This defines a 3-dimensional cochain $\bar{\alpha}_f \in C^3(N;\theta_3^h)$. To show that $\bar{\alpha}_f$ is a cocycle, observe that $\delta\bar{\alpha}_f \mid \tau^4 = \alpha_f \mid \partial\tau^4$, for τ any 4-simplex in N. But $f^{-1}(\tau^4)$ is a homotopy 4-cell which $f^{-1}(\partial\tau^4)$ bounds. Hence $f^{-1}(\partial\tau^4)$ is trivial in θ_3^h, so $\bar{\alpha}_f \mid \partial\tau^4 = 0$. Hence $\delta\bar{\alpha}_f$ is trivial and $\bar{\alpha}_f$ maps to an element α_f in $H^3(N;\theta_3^h)$.

Suppose $f\colon M \to N$ is a smooth CE map of closed n-manifolds, $n \geq 5$, for which α_f is trivial in $H^3(N;\theta_3^h)$. Then $\bar{\alpha}_f = \delta\beta$ for $\beta \in C^2(N;\theta_3^h)$. We now show how to alter f so that $\bar{\alpha}_f = 0$. For simplicity assume β is a cochain which assigns a homotopy 3-sphere H to a single 2-simplex τ in N. Then $\bar{\alpha}_f$ is trivial except on those 3-simplices with τ as a face, to which $\bar{\alpha}_f$ associates H.

Let $(L,\partial L)$ be a properly imbedded $(n-2)$cell in $(N \times I, N \times 0)$ which intersects transversally and exactly once every 3-simplex in $N \times I$ which has $\beta \times 0$ as a face. Furthermore assume L is transverse to $f \times id_I\colon M \times I \to N \times I$. Then there are tubular neighborhoods of $f^{-1}(L)$ and L such that, for appropriate trivializations, f restricts to $[f \mid f^{-1}(L)] \times id_{D^3}\colon f^{-1}(L) \times D^3 \to L \times D^3$. Construct a new manifold W as follows. Remove the tubular neighborhood $f^{-1}(L) \times D^3$ from $M \times I$ and replace by $f^{-1}(L) \times H_0$, where H_0 denotes H with an open 3-ball removed. $f^{-1}(L) \times H_0$ is glued onto $f^{-1}(L) \times \partial D^3$ by the obvious identification of ∂H_0 with ∂D^3. We will show in a moment that there is a smooth CE orientation reversing map $g\colon H_0 \to D^3$ which is the natural identification over ∂D^3. Given g, we can construct a smooth CE map $W \to M \times I$ by extending the natural identification away from $f^{-1}(L) \times H_0$ to $id_{f^{-1}(L)} \times g\colon f^{-1}(L) \times H_0 \to f^{-1}(L) \times D^3$. The composition of this map with f then gives a smooth CE map $\bar{f}\colon W \to N \times I$.

Notice that $\bar{f}^{-1}(\sigma \times 0)$ is $f^{-1}(\sigma \times 0)\#(-H)$ for any 3-simplex σ in N with β as a face. Thus $\bar{f} \mid \bar{f}^{-1}(N \times 0) \to N \times 0$ is a smooth CE map with obstruction trivial in $C^3(N;\theta_3^h)$. But $\bar{f}^{-1}(N \times 0)$ is s-cobordant to M via W. Moreover, the resulting diffeomorphism $h\colon M \to \bar{f}^{-1}(N \times 0)$ is the natural identification outside the star of β. By Lemma 1.1, f is smoothly homotopic through CE maps to $\bar{f}h$, for which the obstruction vanishes in $C^3(N;\theta_3^h)$. Hence we may alter f to make $\overline{\alpha}_f$ trivial.

It remains to construct $g: H_0 \to D^3$. Begin with an orientation reversing homotopy equivalence $\overline{g}: H \to S^3$, and let $q \in S^3$ be a regular point. By a standard argument, $\overline{g}^{-1}(q)$ can be congealed into a single point p by a homotopy of $\overline{g}$. Then $\overline{g}$ is a diffeomorphism over a neighborhood of q. Remove a small open ball around q and also remove the ball's inverse image around p. Then $\overline{g}$ restricts to a homotopy equivalence $\overline{g}_0: H_0 \to B^3$ on the complement, such that $\overline{g}_0$ is a diffeomorphism over a neighborhood of ∂B^3. Then the required g is the composition of $\overline{g}_0$ with $f_r: B^3 \to B^3$, where r is chosen so large that it encloses all of those points over which $\overline{g}_0$ fails to be a diffeomorphism.

§4. The theorem and its proof

Theorem 4.1. *Let $f: M \to N$ be a smooth CE map of closed n-manifolds, $n \geq 6$. Then there is an obstruction $\alpha \in H^3(N;\theta_3^h)$ which vanishes if and only if, given any $\epsilon > 0$, there is a smooth ϵ-homotopy of f through CE maps to a smooth homeomorphism.*

Proof: We have defined a cocycle $\overline{\alpha}_f \in C^3(N;\theta_3^h)$ which clearly vanishes if f is a smooth homeomorphism. Suppose $F: M \times I \to N \times I$ is the smooth level-preserving CE map given by a smooth homotopy through CE maps of $F|M \times 0 = f_0$ to $F|M \times 1 = f_1$. Then the obstruction $\alpha_F \in H^3(N \times I, \theta_3^h)$ restricts to the respective obstructions α_{f_0} and α_{f_1} in $H^3(N;\theta_3^h)$. Since the inclusions of the ends induce isomorphisms on cohomology, α_{f_0}, α_F and α_{f_1} are either all trivial or all non-trivial. This proves that if f can be altered to a homeomorphism, then the obstruction is trivial.

On the other hand, suppose α_f is trivial. In §3 we showed that f may be altered so that $\overline{\alpha}_f$ is trivial, so assume $\overline{\alpha}_f = 0$ in $C^3(N;\theta_3^h)$. Thus the inverse image of a 3-simplex σ is a homotopy 3-cell H h-cobordant to B^3. There are tubular neighborhoods of $H \times 0$ and $\sigma \times 0$ in $M \times I$ and $N \times I$ respectively which can be trivialized so that $f \times id_I$ restricts to the map $[f|H] \times id: H \times \mathrm{cone}(B^{n-3}) \to \sigma \times \mathrm{cone}(B^{n-3})$. We will show in a moment that there is a smooth CE map $g: V \to \sigma \times I$ such that $g|g^{-1}(\sigma \times 0)$ is $f|H$ and g is a homeomorphism over $(\partial\sigma \times I) \cup (\sigma \times 1)$. Given this fact, let W be obtained from $M \times I$ by removing $H \times \text{cone}(B^{n-3})$ and replacing by $(V \times B^{n-3}) \underset{h}{\cup} (\sigma \times \text{cone}(B^{n-3}))$ where h attaches $V \times B^{n-3}$ to $\sigma \times B^{n-3}$ by $g|g^{-1}(\sigma \times 1)$. Then there is a smooth CE map $\bar{f}: W \to N \times I$ which is the natural identification over the complement of $\sigma \times \text{cone}(B^{n-3})$, is $g \times id_{B^{n-3}}$ over the "outer half" of $\sigma \times \text{cone}(B^{n-3})$ and is the homeomorphism $[g|g^{-1}(\sigma \times 1)] \times id$ over the "inner half" of $\sigma \times \text{cone}(B^{n-3})$. As in §3 it is easy to see that W is actually $M \times I$ and that f is smoothly homotopic through CE maps to $\bar{f}|\bar{f}^{-1}(N \times 1)$. In this manner we alter f so that it is a smooth homeomorphism over all of N_3.

The argument that f can be made a homeomorphism over N_4 proceeds in exactly the same way, using the fact that $\theta_4^h = 0$. The remainder of the proof proceeds as in 2.1.

It remains to show that if V is an h-cobordism from a homotopy 3-cell H to B^3, any smooth CE map $g_0: H \to B^3$ which is a homeomorphism over ∂B^3 extends to a smooth CE map $g: V \to B^3 \times I$ which is a homeomorphism over $(\partial B^3 \times I) \cup (B^3 \times 1)$. Ignoring corners, there is, by Lemma 1.2, a smooth CE extension $\partial V \to \partial B^3 \times I$ of g_0 by a smooth homeomorphism. Let $\bar{g}: V \to B^3 \times I$ be any further extension which preserves a collar factor over a neighborhood of $\partial B^3 \times I$. Regard $B^3 \times I$ as B^4 and let g be the composition of $\bar{g}$ with some $f_r: B^4 \to B^4$, where r is chosen so large that f_r collapses the complement of the above collar of $\partial(B^3 \times I)$ to a point. Then g is the required map.

<u>Remark</u>: Given any element φ in θ_3^h there is a smooth CE map $f: M \to S^3 \times T^2$ realizing φ as the obstruction in $H^3(S^3 \times T^2; \theta_3^h) \cong \theta_3^h$. Indeed, if H is a homotopy 3-sphere representing φ and $g: H \to S^3$ is a smooth CE map as constructed for example, in §3, then $g \times id_{T^2}: H \times T^2 \to S^3 \times T^2$ is the appropriate map.

REFERENCES

[CS] M. Cohen and D. Sullivan, Mappings with contractible point inverses between p.l. manifolds, Notices Amer. Math. Soc. 15 (1968), 186. Abstract #653-363.

[Sc_1] M. Scharlemann, Approximating CAT CE maps by CAT homeomorphisms, to appear in the proceedings of the 1977 Georgia topology conference.

[Sc_2] _______, Transverse Whitehead triangulations, to appear.

[SS] M. Scharlemann and L. C. Siebenmann, The Hauptvermutung for C^∞ homeomorphisms II. Composito Math. 29 (1974), 253-263.

[Si] L. Siebenmann, Approximating cellular maps by homeomorphisms, Topology 11 (1972), 271-294.

[Wh] J.H.C. Whitehead, On C^1 complexes, Ann. Math. 41 (1940), 804-824.